KU-511-254
INSTITUTE
STATISTICS
OXFORD.
WITHDRAWN

ELEMENTS OF PROBABILITY THEORY

ELEMENTS OF PROBABILITY THEORY

BY

L. Z. RUMSHISKII

TRANSLATED FROM THE RUSSIAN

BY

D. M. G. WISHART

Lecturer in Statistics
University of Birmingham

PERGAMON PRESS

OXFORD · LONDON · EDINBURGH · NEW YORK

PARIS · FRANKFURT

Pergamon Press Ltd., Headington Hill Hall, Oxford
4 & 5 Fitzroy Square, London W. 1

Pergamon Press (Scotland) Ltd., 2 & 3 Teviot Place, Edinburgh 1

Pergamon Press Inc., 122 East 55th St., New York 22, N.Y.

Pergamon Press GmbH, Kaiserstrasse 75, Frankfurt-am-Main

First edition 1965

Library of Congress Catalog Card No. 63-22490

A translation of the original volume,
Elementy Teorii Veroyatnostei, Fizmatgiz,
Moscow (1960)

1594

CONTENTS

FOREWORD

This book is intended for colleges of advanced technology in which the advanced mathematics syllabus includes an account of the elements of the theory of probability and mathematical statistics in a course of 30 hours. It is assumed that the student has a knowledge of calculus such as may be obtained from the ordinary syllabus of a technical college. The book introduces the basic ideas and some of the methods of the theory of probability, which are necessary today in several technical fields. The theory of random (stochastic) processes and other special questions requiring a more advanced mathematical preparation have not been included in this book. The more difficult parts of the course, which lie outside the limits of a minimal programme, have been printed in small type and may be omitted at the first reading. The examples in the text play an essential role in the clarification of the basic concepts; the student is recommended to examine them in detail.

The book is based on a course of lectures given by the author over a period of ten years in the Moscow Power Institute.

The author is deeply grateful to A. M. Yaglom for his invaluable advice during work on the manuscript.

The author would also like to take this opportunity to thank R. Ya. Berr, I. A. Brin, M. I. Vishik, S. A. Stebakov, and R. S. Has'minskii, for their useful observations.

TRANSLATOR'S PREFACE

AFTER the manuscript of this translation had been completed it was apparent that the number of exercises was inadequate for a textbook on this subject. I have therefore more than doubled the number of exercises, drawing extensively from *A Guide for Engineers to the Solution of Problems in the Theory of Probability* by Professor A. A. Svenshnikov *et al.* (Leningrad, 1962). I have added a few exercises of my own where the *Guide* did not cover a topic which I thought worth pursuing. In addition, Tables III and IV have been copied from the *Guide*.

The reference on p. 125 replaces a reference to a Russian textbook unavailable in translation. The references to Professor Gnedenko's excellent book have been left, as it is now available in English.

I am indebted to Mr. J. N. Darroch who read the manuscript and made a number of useful suggestions.

D. M. G. WISHART

INTRODUCTION

IN VARIOUS fields of technology and production it is becoming more frequently necessary to deal with mass phenomena. These mass phenomena have their own regularity. Consider for example the process of the automatic machining of components in a production line. The dimensions of the different components will vary about some central value. These fluctuations have a random character, in the sense that the measurement of one component does not allow us to predict precisely the value of the measurement of the next component. However, the distribution of the measurements of a large batch conforms rather precisely to a pattern. Thus, the arithmetic means of the measurements of the components in different batches are approximately the same, and the deflexions of the different values from the mean of the measurements are also found with almost equal frequencies in different batches. A similar situation can be observed in the repeated weighing of a body on an analytical balance. Here, the different results of the measuring process differ one from another; nevertheless the mean value of several measurements is practically invariant, and it is also possible to calculate precisely the frequencies of the different deviations from this mean value. Regularity of this kind does not allow us to predict the result of a single measurement but it does give us a basis for analysing the results of a large number of measurements.

Regularities of this kind–the stability of mean values and the frequencies of occurrence of random fluctuations in repeated experiments–these are the subject matter of the Theory of Probability.

These patterns were first observed in the seventeenth century in connection with games of chance, particularly in games with dice. In repeated throws of a die it was noticed that each number from 1 to 6 appeared approximately equally often, or in other words appeared with a frequency of about $\frac{1}{6}$. With throws of two dice the sum of the two numbers takes all possible whole numbers from 2 to 12 but not now equally often; however, the relative frequencies of these possible values (in a large number of throws) will be near to numbers which can be calculated beforehand according to a simple rule (see p. 24). The establishment of similar rules and the solution of somewhat more complicated problems connected with the throwing of dice played an important role in the early period of the development of probability theory. And even today the basic ideas of the theory (an event and its probability, a random variable, etc.) can be conveniently illustrated by examples with dice. Thus in the throw of a die and the observation of the number falling on the upper face we have a clear illustration of the idea of a trial with a set of equally likely outcomes and of the idea of the simplest events; the number itself serves as a simple example of a random variable–a variable whose values depend on chance.

The simplest schemes, created for the solutions of questions connected with the throwing of dice, have a very limited field of application. Various questions, arising in the development of probability theory in technology and the natural sciences during the nineteenth and twentieth centuries, required the study of random variables of a considerably more complicated nature, in particular the so-called continuous random variables. As examples of such variables we may think of the measurements of machined parts and the results of weighings which we considered above. Continuous random variables and their numerical characteristics will be given fundamental consideration in this book.

We have already indicated our approach to the subject and accordingly in the first chapter we will consider briefly the primary ideas of events and their probabilities.

In the second chapter we will give a detailed discussion of random variables and functions of random variables of the two most important types–discrete and continuous. Our considerations will be limited to the basic case of one-dimensional random variables, and we will deal separately with the discrete and continuous types.

In the third chapter the fundamental numerical characteristics of a random variable will be given, and the simplest properties proved. The most important properties of the mean value of a random variable, and the connection with the so-called law of large numbers, are discussed in the fourth chapter. The fifth chapter is devoted to one of the central questions of probability theory–that of limit theorems–and of the elucidation of the role played by the so-called normal law (in particular for the estimation of mean values).

In the sixth chapter we will consider some applications of the theory to the analysis of numerical data (i.e. the theory of errors of measurement).

Finally, in the seventh chapter, we will examine the important practical question of linear correlation between random variables.

CHAPTER I

EVENTS AND PROBABILITIES

§ 1. EVENTS. RELATIVE FREQUENCY AND PROBABILITY

Let us suppose that we perform an experiment. (A performance of the experiment will be called *a trial.*) Any possible outcome of a performance of this experiment will be called *an event*. We will suppose that we may perform the experiment an infinite number of times.

EXAMPLE. Trial–the throw of a die: events–the occurrence of a 6; the occurrence of an even number of points.

EXAMPLE. Trial–the weighing of an object in an analytical balance: event–the error of measurement (i.e. the difference between the result of the weighing and the true weight of the body) does not exceed a previously given number.

If in n trials a given event occurs m times, the relative frequency of the event is the ratio m/n. Experience shows that for repeated trials this relative frequency possesses a definite stability: if, for example, in a large series of n trials the relative frequency is $m/n = 0{\cdot}2$, then in another long series of n' trials the relative frequency m'/n' will be near 0·2. Thus in different sufficiently long series of trials the relative frequencies of an event appear to be grouped near some constant number (different numbers, ofcourse, for different events). For example, if a die is a perfect cube, and made of a homogeneous material (an "unbiased" die) then the numbers 1, 2, 3, 4, 5 and 6 will each appear with relative frequency near to $\frac{1}{6}$.

Because of the nature of these experiments, the relative frequencies of the events are stable. For example, in the throwing of an unbiased die the approximate equality of the relative frequencies with which the six faces appear is explained by its symmetry, giving the same possibility of occurrence to each number from 1 to 6.

Thus *we assign to an event a number called the probability of the event. This measures the degree of possibility of occurrence of the event, in the sense that the relative frequencies of this event obtained from repetitions of the experiment are grouped near this number.* As with the relative frequency of an event, its probability must be dimensionless, a constant lying between 0 and 1. However, we emphasize the difference that, whereas the relative frequency still depends on the carrying out of trials, the probability of an event is connected only with the event itself (as a possible outcome of a given experiment).

Thus probability is the first basic idea, and in general it is impossible to define it more simply. As we shall show in the following section, we can calculate the probabilities directly only in certain very simple schemes; the analysis of these simple experiments will allow us to establish the basic properties of probability which we shall need for the further development of the theory.

§ 2. THE CLASSICAL DEFINITION OF PROBABILITY

Let us first of all agree on some notation. Events are called *mutually exclusive* if they cannot occur simultaneously. A collection of events form a *partition* if at each trial one and only one of the events must occur; i.e. if the events are pair-wise mutually exclusive and if only one of them occurs.

In this section we restrict our attention to *trials with equally likely outcomes*; for example we shall consider the throwing of an unbiased die with possible outcomes 1, 2, 3, 4, 5 and 6.† In other words we shall study trials whose possible outcomes can be represented by a partition of equally likely events; in these circumstances the events will be known as "cases".

If the partition consists of N equally likely cases, then each case will have probability equal to $1/N$. This accords with the fact that in a large number of trials equally likely cases occur approximately

† We shall not try to define the idea of "equally likely outcomes" in terms of any simpler concepts. It is usually based on some consideration of symmetry, as in the example of the die, and connected in practice with the approximate equality of the relative frequencies of all the outcomes in a large number of trials. We remark that everywhere in this section we shall assume that we are dealing with a finite number of cases.

the same number of times, i.e. they have relative frequencies near to $1/N$. For example, in throwing an unbiased die the cases are the appearance of 1, 2, 3, 4, 5, and 6 points, and these form a partition; each case will have probability equal to $\frac{1}{6}$.

Let us consider now a compound event A comprising M cases. The probability of the event A is defined to be M/N. For example, the probability that an even number appears as a result of throwing our unbiased die is equal to $\frac{3}{6}$, because among the six equally likely cases only three of them (2, 4, 6) represent the appearance of an even number.

The probability of an event A will be denoted by the symbol $\mathbf{P}\{A\}$.

The expression

$$\mathbf{P}\{A\} = \frac{M}{N} \tag{1.1}$$

expresses the so-called classical definition of probability: *if the outcomes of a trial can be represented in the form of a partition into N equally likely cases, and if the event A occurs only in M cases, then the probability of the event A is equal to M/N; i.e. the ratio of the number of cases "favourable" to the event to the total number of cases.*

EXAMPLE. If we toss two coins, tails can appear twice, once, or not at all: we wish to determine the probabilities of these three events (assuming that for each coin the appearance and non-appearance of tails are equally likely).

Now the three events themselves form a partition, since they are evidently mutually exclusive. But they are not equally likely, and in order to apply the classical definition of probability we must represent the possible outcomes of a trial as a partition of equally likely events. By considerations of symmetry this can be done in the following way:

1st Coin	2nd Coin
Tail	Tail
Tail	Head
Head	Tail
Head	Head

It is natural to consider as equally likely the four outcomes enumerated here, and moreover, they again form a partition. We can therefore now apply the classical definition of probability and deduce:

$$\mathbf{P}\{\text{two tails appear}\} = \tfrac{1}{4},$$

$$\mathbf{P}\{\text{one tail appears}\} = \tfrac{2}{4} = \tfrac{1}{2},$$

$$\mathbf{P}\{\text{no tail appears}\} = \tfrac{1}{4}.$$

We emphasise again that the classical definition of probability rests essentially on the assumption that the outcomes of a trial are all equally possible. All problems to which the classical definition (1.1) applies can be formulated in terms of the following simple scheme–the scheme of random sampling: one element is drawn at random from a set of N elements (objects, phenomena, etc.) so that each element has the same possibility of being drawn; the event A occurs when we draw an element having some given mark, which mark is possessed by precisely M of the N elements in the set.

The simplest realisation of this scheme is the following: in an urn suppose that there are N balls, identical to the touch, and suppose that M of them are white and $N-M$ are not white; a trial consists in drawing one ball at random from the urn; the event A consists in drawing a white ball. Under these conditions, the probability of drawing a white ball is equal to M/N.

§ 3. FUNDAMENTAL PROPERTIES OF PROBABILITIES. RULE FOR THE ADDITION OF PROBABILITIES

The analysis which we have given of the classical definition of probability suggests the following fundamental properties of probabilities in general.

(1) The probability of an event is a non-negative number:

$$\mathbf{P}\{A\} \geqq 0. \tag{1.2}$$

(2) The certain event (the event which, when we perform the experiment, occurs without fail) has probability one:

$$\mathbf{P}\{\text{the certain event}\} = 1. \tag{1.3}$$

(3) The probabilities of events obey the rule of addition: if the event C occurs when one of two mutually exclusive events A, B occurs, then the probability of the event C is equal to the sum of the probabilities of the events A and B.

We express this rule in the form:

$$\mathbf{P}\{A \text{ or } B\} = \mathbf{P}\{A\} + \mathbf{P}\{B\} \quad \text{(for mutually exclusive } A \text{ and } B). \tag{1.4}$$

This relation is also known as *the additivity property for probabilities.*

The first two conditions follow immediately from (1.1), because in this case $M \geqq 0$, $N > 0$ and for the certain event $M = N$ (all outcomes of a trial are favourable to the certain event). The third condition is verified in the following way for the scheme of random sampling. Let us suppose that there are N balls in an urn, of which K are red, L are blue, and the remainder are white; a trial will consist of drawing one ball at random from the urn; the event A will occur if a red ball is drawn, and the event B will occur if a blue ball is drawn. Then the event (A or B) consists in the drawing of a coloured ball (either red or blue). The direct calculation of the probabilities acording to (1.1) gives:

$$\mathbf{P}\{A\} = \frac{K}{N}; \quad \mathbf{P}\{B\} = \frac{L}{N};$$

$$\mathbf{P}\{A \text{ or } B\} = \frac{K+L}{N},$$

which agrees with (1.4).

In order to apply the theory of probability it is extremely important that these properties should be valid not only for the scheme of random sampling but also for any system of events. We can base such an assertion on the following argument. We recall that the general idea of probability arose from the observed stability of the relative frequency of an event. It is natural therefore to require that the basic properties of probabilities should coincide with the corresponding properties of relative frequencies. Now properties 1, 2 and 3 are easily verified for relative frequencies.

(1)† A relative frequency m/n cannot be negative since $m \geqq 0$, $n > 0$.

(2)† The certain event occurs with each trial and therefore its relative frequency is equal to $n/n = 1$.

(3)† If the events are mutually exclusive then the event (A or B) occurs as often as one or other of them occurs (thus the number of drawings of coloured balls from the urn described above is equal to the number of drawings of red balls and blue balls). The relative frequency of the event (A or B) is therefore equal to the sum of the relative frequencies of the events A and B.

From these considerations *we take the above three statements as the basic properties of probabilities for an arbitrary system of events.*†

Note on the subject matter of probability theory

The theory of probability does not study the "true nature" of the different events but only quantitative relationships between their probabilities. A typical question for the theory and its applications is the following: we are given some collection of simple events and their probabilities: we are required to find the probabilities of some other events connected with the simple events in a well-defined way. For example, if at each toss of a coin the probability of its falling tails is $\frac{1}{2}$, what is the probability of getting not less than 50 tails in 100 tosses? The solutions to these questions follow from the rules of composition (one of which – the additivity property – we have already introduced). For all applications it is quite inessential exactly how we define the probabilities of the simple events. It is important only that for a sufficiently large number of trials the relative frequencies of these events should be near to their probabilities: then this will also be true for the compound events in which we are interested (whose probabilities we calculate by means of our rules). The rules of composition which we adopt for the theory of probability are consistent with this basic requirement.

† We remark here that on the basis of these and some other properties it is possible to construct an axiomatic system for probability theory: this construction was successfully carried out by the well-known Soviet mathematician Academician A. N. Kolmogorov (for the Anglo-American reader see A. N. Kolmogorov, *Foundations of Probability*, Chelsea, 1950). Consideration of this question lies outside the framework of this book.

Consequences of the fundamental properties of probabilities

(1) If the events $A_1, A_2, \ldots, A_n$ are mutually exclusive then

$$\mathbf{P}\{A_1 \text{ or } A_2 \text{ or} \ldots \text{or } A_n\} = \mathbf{P}\{A_1\} + \mathbf{P}\{A_2\} + \cdots + \mathbf{P}\{A_n\}. \quad (1.5)$$

We derive (1.5) easily from (1.4) using the principle of induction.

(2) If the mutually exclusive events $A_1, A_2, \ldots, A_n$ form a partition, then the sum of their probabilities is equal to one.

In fact, by the definition of a partition, the event (A_1 or A_2 or ... or A_n) is certain and therefore

$$\mathbf{P}\{A_1 \text{ or } A_2 \text{ or} \ldots \text{or } A_n\} = 1.$$

Applying (1.5) to the left side of this relation, we obtain

$$\mathbf{P}\{A_1\} + \mathbf{P}\{A_2\} + \cdots + \mathbf{P}\{A_n\} = 1. \quad (1.6)$$

Particular interest attaches to the case when the partition consists of only two events. The occurrence of one of them is thus equivalent to the non-occurrence of the other. We will call such events *complementary*. If the event A is one of a pair of complementary events, then we shall denote the other by $\bar{A}$ (to be read "not A"). The sum of the probabilities of two complementary events is equal to one:

$$\mathbf{P}\{A\} + \mathbf{P}\{\bar{A}\} = 1. \quad (1.7)$$

This follows directly from (1.6).

Thus, if the probability of some event A is known, the probability of its complement $\bar{A}$ is given by

$$\mathbf{P}\{\bar{A}\} = 1 - \mathbf{P}\{A\}.$$

As an example of complementary events we can take the appearance of heads or tails when we toss a coin. If these events are equally likely then each will have the probability $\frac{1}{2}$.

The impossible event

An event which cannot occur as the outcome of a trial will be called an *impossible event*. Examples: drawing a white ball from an urn containing no white balls; or obtaining a negative result from the weighing of a body. An impossible event can be thought of as the complement of some certain event associated with the experiment; therefore the probability of an impossible event is equal to zero.

We remark that by the classical definition the probability of an event is equal to zero if and only if there is no possible outcome of a trial which is favourable to this event (i.e. $M = 0$). We shall see later in the study of continuous random variables that an event whose probability is zero is not necessarily impossible.

The relation between the classical definition and the fundamental properties

As we emphasised above, the classical definition of probability is concerned with the case where the outcomes of a trial can be represented as a partition into equally likely events. It is worth noticing that in this case the classical formula (1.1) can be deduced from the three fundamental properties of probabilities.

We prove the following assertion: If the elementary outcomes of a trial form a partition of equally likely events, then the probability of a compound event is given by the classical formula (1.1).

To prove this we denote the elementary outcomes by $E_1, E_2, \dots, E_N$ and we write their common probability as $p = \mathbf{P}\{E_k\}$ $(k = 1, 2, \dots, N)$. From (1.6) we have

$$\mathbf{P}\{E_1\} + \mathbf{P}\{E_2\} + \cdots + \mathbf{P}\{E_N\} = 1,$$

so that $Np = 1$. A is a compound event for which M given elementary outcomes ($E_1, E_2, \dots, E_M$, say) are favourable. Then A is the compound event (E_1 or E_2 or ... or E_M) and applying (1.5) we obtain

$$\mathbf{P}\{A\} = \mathbf{P}\{E_1\} + \mathbf{P}\{E_2\} + \cdots + \mathbf{P}\{E_M\}$$

so that

$$\mathbf{P}\{A\} = Mp = \frac{M}{N}.$$

§ 4. THE INTERSECTION OF EVENTS. INDEPENDENT EVENTS

The event consisting of the simultaneous occurrence of the events A and B is called the *intersection* of the events A and B. We shall denote this event by (A and B).

EXAMPLE. We pick one number at random from the numbers 1, 2, ..., 100. The event A consists of the numbers divisible by 3 and the event B consists of the numbers divisible by 4. Then the event (A and B) consists of the numbers divisible by both 3 and 4, i.e. the numbers divisible by 12. It is easy to see that

$$\mathbf{P}\{A\} = \frac{33}{100}; \quad \mathbf{P}\{B\} = \frac{25}{100}; \quad \mathbf{P}\{A \text{ and } B\} = \frac{8}{100}$$

since among the first 100 numbers there are 33 numbers divisible by 3, 25 numbers divisible by 4, and 8 numbers divisible by 12.

The simplest relationship between the probabilities of the events A and B, and the probability of the intersection (A and B) takes place when the events A and B are *independent*. We shall exemplify the idea of independence first of all for the random sampling scheme. Let us suppose that we have two urns containing balls, and that we draw one ball at random from each urn. Let the event A consist in drawing a white ball from the first urn, and the event B in drawing a white ball from the second urn. These events are essentially independent in the sense that the colour of the ball drawn from the first urn cannot influence the colour of the ball drawn from the second urn. We require the probability of the intersection (A and B), i.e. the probability of drawing white balls from both urns. Let us denote the number of balls in the first and second urns by N_1 and N_2 respectively, and the number of white balls in them by M_1 and M_2 respectively. Then

$$\mathbf{P}\{A\} = \frac{M_1}{N_1} \quad \text{and} \quad \mathbf{P}\{B\} = \frac{M_2}{N_2}.$$

Because each of the N_1 possible outcomes of the first draw can be combined with each of the N_2 possible outcomes of the second draw, it follows that the total number of possible outcomes is equal to $N_1 N_2$. The drawing of two white balls occurs in only $M_1 M_2$ of these cases. Consequently the probability of the intersection is given by

$$\mathbf{P}\{A \text{ and } B\} = \frac{M_1 M_2}{N_1 N_2},$$

or

$$\mathbf{P}\{A \text{ and } B\} = \mathbf{P}\{A\}\,\mathbf{P}\{B\}. \tag{1.8}$$

This expresses the multiplication rule for the probabilities of independent events.

We recall that this relation has been demonstrated only for the particular scheme of random sampling. In general the independence or non-independence of the events in which we are interested will have to be included among the assumptions. The simplicity of (1.8) makes independence an important instrument in calculations.

DEFINITION. *Two events A and B are called independent if* (1.8) *holds; i.e., if the probability of their intersection is equal to the product of their probabilities.*

We remark that from the independence of the events A and B we can deduce the pair-wise independence of the events $\bar{A}$ and B, A and $\bar{B}$, $\bar{A}$ and $\bar{B}$. We leave this deduction to the reader as an easy exercise in the formal handling of probabilities (Exercise 8, p. 18).

The definition which we have given above for the independence of two events can be extended to a larger number of events: *the events $A_1, A_2, \ldots, A_n$ are called mutually independent if the probability of the intersection of any* 2, 3, ..., n *of them is equal to the product of the corresponding probabilities.* For example, the three events A, B, C are mutually independent if four relations hold:

$$\mathbf{P}\{A \text{ and } B\} = \mathbf{P}\{A\}\,\mathbf{P}\{B\}; \quad \mathbf{P}\{A \text{ and } C\} = \mathbf{P}\{A\}\,\mathbf{P}\{C\};$$

$$\mathbf{P}\{B \text{ and } C\} = \mathbf{P}\{B\}\,\mathbf{P}\{C\};$$

$$\mathbf{P}\{A \text{ and } B \text{ and } C\} = \mathbf{P}\{A\}\,\mathbf{P}\{B\}\,\mathbf{P}\{C\}. \tag{1.9}$$

The following example is useful in showing events which are pair-wise independent but not mutually independent. Consider an urn containing four balls numbered 1, 2, 3, and 123, and suppose a trial to consist in drawing one ball at random. Let A, B, C be the events associated with the appearance of the digits 1, 2, 3 respectively. These events A, B, C are pair-wise independent because

$$\mathbf{P}\{A\} = \mathbf{P}\{B\} = \mathbf{P}\{C\} = \tfrac{2}{4} = \tfrac{1}{2};$$

$$\mathbf{P}\{A \text{ and } B\} = \mathbf{P}\{B \text{ and } C\} = \mathbf{P}\{C \text{ and } A\} = \tfrac{1}{4} = \tfrac{1}{2} \times \tfrac{1}{2}.$$

But these events are not mutually independent, because

$$\mathbf{P}\{A \text{ and } B \text{ and } C\} = \tfrac{1}{4} \neq \tfrac{1}{2} \times \tfrac{1}{2} \times \tfrac{1}{2}.$$

We remark also that the mutual independence of the events A, B, and C requires the simultaneous satisfaction of all the relations (1.9). For example, if we have an urn containing 8 balls numbered 1, 2, 3, 12, 13, 20, 30, 123, and if the events A, B, C, have the same sense as in the previous example, then

$$\mathbf{P}\{A\} = \mathbf{P}\{B\} = \mathbf{P}\{C\} = \tfrac{4}{8} = \tfrac{1}{2}; \quad \mathbf{P}\{A \text{ and } B \text{ and } C\} = \tfrac{1}{8} = \tfrac{1}{2} \times \tfrac{1}{2} \times \tfrac{1}{2};$$

and further

$$\mathbf{P}\{A \text{ and } B\} = \tfrac{2}{8} = \tfrac{1}{2} \times \tfrac{1}{2}; \quad \mathbf{P}\{A \text{ and } C\} = \tfrac{2}{8} = \tfrac{1}{2} \times \tfrac{1}{2};$$

but

$$\mathbf{P}\{B \text{ and } C\} = \tfrac{1}{8} \neq \mathbf{P}\{B\}\,\mathbf{P}\{C\}.$$

Generalisation of the additivity property

If the events A and B are independent, then

$$\mathbf{P}\{A \text{ or } B\} = \mathbf{P}\{A\} + \mathbf{P}\{B\} - \mathbf{P}\{A\}\,\mathbf{P}\{B\}. \qquad (1.10)$$

To prove this relation, we note, first of all, that the events (A or B) and ($\bar{A}$ and $\bar{B}$) are complementary (for if one or other of the events A or B occurs then the corresponding complementary event does not occur, and so the intersection of the complementary events $\bar{A}$ and $\bar{B}$ cannot occur). The application of (1.7) and (1.8) yields

$$\mathbf{P}\{A \text{ or } B\} = 1 - \mathbf{P}\{\bar{A} \text{ and } \bar{B}\} = 1 - \mathbf{P}\{\bar{A}\}\,\mathbf{P}\{\bar{B}\}$$

$$= 1 - (1 - \mathbf{P}\{A\})\,(1 - \mathbf{P}\{B\}) = \mathbf{P}\{A\} + \mathbf{P}\{B\} - \mathbf{P}\{A\}\,\mathbf{P}\{B\},$$

which was to be proved.

We state without proof the general addition rule for probabilities:

$$\mathbf{P}\{A \text{ or } B\} = \mathbf{P}\{A\} + \mathbf{P}\{B\} - \mathbf{P}\{A \text{ and } B\}. \qquad (1.11)$$

(The events A and B are not necessarily independent here.)

EXERCISES

(1) Prove the relation (1.11) using the classical definition of probabilities.

(2) Give a geometrical interpretation of (1.11) in terms of the following experiment. Consider the unit square and let a trial consist of choosing a point at random. Assume that the probability of choosing a point in a given set A within the square is equal to the area of this set (see Fig. 1).

EXAMPLE 1. Two riflemen shoot independently at the same target. They fire one shot each; the probability of a hit for the first riflemen is $\mathbf{P}\{A\} = 0{\cdot}9$, and for the second riflemen is $\mathbf{P}\{B\} = 0{\cdot}8$. From (1.10) we determine the probability of at least one hit on the target:

$$\mathbf{P}\{A \text{ or } B\} = 0{\cdot}9 + 0{\cdot}8 - (0{\cdot}9)\,(0{\cdot}8) = 0{\cdot}98.$$

EXAMPLE 2. n riflemen shoot independently at the same target. We assume that the probability of a hit is equal to p for each rifleman. The problem is to determine the number of riflemen which are necessary to make the probability of a hit is not less than P.

We seek a relation between p and P. The probability that a given rifleman will not hit the target is equal to $1 - p$; the probability that none of the riflemen will hit the target is therefore $(1 - p)^n$.

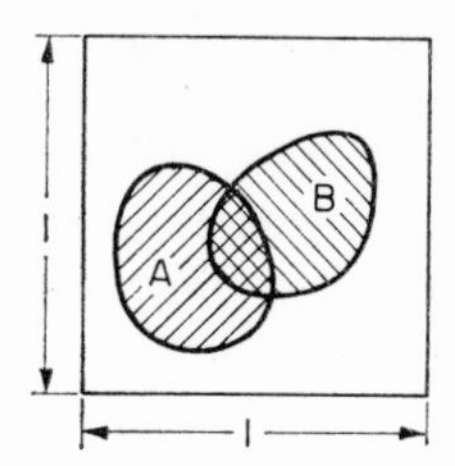

FIG. 1

The events "no rifleman hits the target" and "at least one rifleman hits the target" (i.e. "the target receives a hit") are complementary, so that the probability of a hit is $1 - (1 - p)^n$. The required relation is

$$1 - (1 - p)^n \geqq P.$$

From this we derive the inequality

$$n \geqq \frac{\lg(1 - P)}{\lg(1 - p)}.$$

For example, if in firing at an aeroplane the probability of a hit is $p = 0{\cdot}004$, we seek the number of shots so that the probability of a hit is at least $P = 0{\cdot}98$. We obtain

$$n \geqq \frac{\lg 0{\cdot}02}{\lg 0{\cdot}996}, \quad \text{or} \quad n \geqq 976 \text{ shots}.$$

§ 5. CONDITIONAL PROBABILITIES. THE GENERAL RULE FOR THE MULTIPLICATION OF PROBABILITIES. THE FORMULA OF TOTAL PROBABILITY

The aim of the present paragraph is to generalise the multiplication rule for probabilities (1.8) to events which are not independent. As in § 4 we begin with the classical scheme of random sampling. Let us consider the experiment of drawing at random one ball from an urn containing N balls. The balls are identical in all respects except that some of the balls are coloured (the remainder are white), and some of the balls have a pattern drawn on the surface (and the remainder have not). We will suppose that

K balls are coloured ($N - K$ are white)

L balls are patterned ($N - L$ are not)

M balls are both coloured and patterned.

Let the event A be the appearance of a coloured ball, and the event B be the appearance of a patterned ball. The intersection of the events A and B denotes the appearance of a ball which is both coloured and patterned. The probabilities of these events are

$$\mathbf{P}\{A\} = \frac{K}{N}; \quad \mathbf{P}\{B\} = \frac{L}{N}; \quad \mathbf{P}\{A \text{ and } B\} = \frac{M}{N}.$$

By analogy with (1.8) we try to relate the probability of the event (A and B) to the probability of the event A. We obtain

$$\frac{M}{N} = \frac{K}{N} \cdot \frac{M}{K}. \tag{1.12}$$

The ratio M/K of the number of coloured balls with patterns to the number of all coloured balls has the character of a probability. In fact it gives the probability of drawing a patterned ball on condition that the draw is made only from those balls which are coloured (this follows from the observation that of the K coloured balls, only M have patterns). Such a probability will be called *the conditional probability of the event B given that the event A has* *been realised.* We will denote this conditional probability by $\mathbf{P}\{B|A\}$: in our example

$$\mathbf{P}\{B|A\} = \frac{M}{K}.$$

Now we can write (1.12) in the form

$$\mathbf{P}\{A \text{ and } B\} = \mathbf{P}\{A\}\,\mathbf{P}\{B|A\}. \tag{1.13}$$

This relation expresses the general rule for the multiplication of probabilities: *the probability of the intersection of two events is equal to the product of the probability of one of them and the conditional probability of the other.*

We derived (1.13) for the classical scheme. We turn now to the general case of arbitrary events A and B. Formula (1.13) serves here as a definition of conditional probability. In fact, *we define the conditional probability of the event B given the event A by the relation*

$$\mathbf{P}\{B|A\} = \frac{\mathbf{P}\{A \text{ and } B\}}{P\{A\}} \tag{1.14}$$

provided $\mathbf{P}\{A\} \neq 0$. In precisely the same way we can introduce the conditional probability of the event A given that the event B has occurred:

$$\mathbf{P}\{A|B\} = \frac{\mathbf{P}\{A \text{ and } B\}}{\mathbf{P}\{B\}} \tag{1.15}$$

provided $\mathbf{P}\{B\} \neq 0$. It is easy to verify that conditional probabilities possess all the basic properties of probabilities.

We can generalise (1.13) to the case of a larger number of events. Thus, for three events A, B, and C

$$\begin{aligned}\mathbf{P}\{A \text{ and } B \text{ and } C\} &= \mathbf{P}\{A \text{ and } B\}\,\mathbf{P}\{C|A \text{ and } B\}\\ &= P\{A\}\,\mathbf{P}\{B|A\}\,\mathbf{P}\{C|A \text{ and } B\}.\end{aligned}$$

The idea of conditional probability allows us to give a new interpretation to the independence of events. If the events A and B are independent, then from (1.8), (1.14) and (1.15) it follows that

$$\mathbf{P}\{B|A\} = \frac{\mathbf{P}\{A\}\,\mathbf{P}\{B\}}{\mathbf{P}\{A\}} = \mathbf{P}\{B\},$$

$$\mathbf{P}\{A|B\} = \frac{\mathbf{P}\{A\}\,\mathbf{P}\{B\}}{\mathbf{P}\{B\}} = \mathbf{P}\{A\}$$

so that the conditional and unconditional probabilities of these events are equal.

It is clear also that if conversely $\mathbf{P}\{B|A\} = \mathbf{P}\{B\}$, then (1.13) becomes (1.8).

Thus the independence of the events A and B implies that the probability of the event B (or A) does not depend on the realisation of the event A (or B).†

It follows immediately from this interpretation that the certain event and an arbitrary event A are independent.

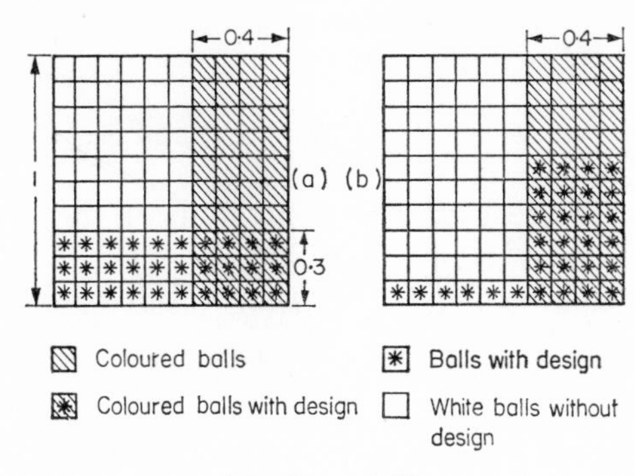

FIG. 2a and 2b

In the above example the independence of the events A and B was seen to be equivalent to the relation

$$\frac{M}{K} = \frac{L}{N};$$

i.e. the ratio of the number of coloured balls with a pattern to the number of coloured balls is equal to the ratio of the number of balls with a pattern to the total number of balls in the urn. We represent this diagrammatically in Fig. 2a, where $\mathbf{P}\{A\} = 0{\cdot}4$, $\mathbf{P}\{B\} = \mathbf{P}\{B|A\} = 0{\cdot}3$. Figure 2b is given for comparison. In this case

$$\mathbf{P}\{A\} = 0{\cdot}4,\ \mathbf{P}\{B\} = \frac{24+6}{100} = 0{\cdot}3,\ \text{but } \mathbf{P}\{B|A\} = \frac{24}{40} = 0{\cdot}6.$$

† We recall that the probability of an event B is always associated with the performance of an experiment. If the experiment is altered by the additional condition that some other event A be realised, then the probability of the event B need not remain the same.

The formula of total probability

THEOREM. *If the events $H_1, \ldots, H_n$ form a partition, then for any event A,*

$$\mathbf{P}\{A\} = \mathbf{P}\{H_1\}\,\mathbf{P}\{A|H_1\} + \mathbf{P}\{H_2\}\,\mathbf{P}\{A|H_2\} + \cdots + \mathbf{P}\{H_n\}\,\mathbf{P}\{A|H_n\}. \qquad (1.16)$$

Proof: From the conditions of the theorem we can write the event A in the form

$$(H_1 \text{ or } H_2 \text{ or} \ldots \text{or } H_n) \text{ and } A.$$

This event will be realised if and only if one of the joint events

$$(H_1 \text{ and } A) \text{ or } (H_2 \text{ and } A) \text{ or} \ldots \text{or } (H_n \text{ and } A)$$

is realised. Since $H_1, H_2, \ldots, H_n$ are mutually exclusive we may add the probabilities and obtain

$$\mathbf{P}\{A\} = \mathbf{P}\{(H_1 \text{ or } H_2 \text{ or} \ldots \text{or } H_n) \text{ and } A\}$$

$$= \mathbf{P}\{H_1 \text{ and } A\} + \mathbf{P}\{H_2 \text{ and } A\} + \cdots + \mathbf{P}\{H_n \text{ and } A\}. \qquad (1.17)$$

If we apply now the general rule for the multiplication of probabilities, $\mathbf{P}\{H_1 \text{ and } A\} = \mathbf{P}\{H_1\}\,\mathbf{P}\{A|H_1\}$, etc., we complete the proof.

In particular we have always the relation

$$\mathbf{P}\{B\} = \mathbf{P}\{A\}\,\mathbf{P}\{B|A\} + \mathbf{P}\{\bar{A}\}\,\mathbf{P}\{B|\bar{A}\}, \qquad (1.18)$$

since the complementary events A and $\bar{A}$ are mutually exclusive and form a partition.

EXAMPLE 1. We consider our urn containing N balls, M of which are white. A trial consists in drawing two balls from the urn consecutively without replacement. The event A occurs if a white ball appears at the first draw: the event B occurs if a white ball appears at the second draw.

Clearly

$$\mathbf{P}\{A\} = \frac{M}{N}; \quad \mathbf{P}\{\bar{A}\} = \frac{M-N}{N};$$

$$\mathbf{P}\{B|A\} = \frac{M-1}{N-1}; \quad \mathbf{P}\{B|\bar{A}\} = \frac{M}{N-1}.$$

We calculate $\mathbf{P}\{B\}$ using (1.18):

$$\mathbf{P}\{B\} = \frac{M}{N} \cdot \frac{M-1}{N-1} + \frac{N-M}{N} \cdot \frac{M}{N-1} = \frac{M}{N}.$$

Thus the probability that a white ball occurs at the second draw is equal to the probability that a white ball occurs at the first draw.

EXAMPLE 2. Slots with coordinates (x_k, y_i) $(k = 1, 2, \ldots, n;$ $i = 1, 2, \ldots, m)$ are made in a board. A small ball thrown at this board will finally get into one of these slots. The probabilities of getting into each slot are given in the table:

$y \backslash x$	x_1	x_2	$\ldots$	x_n
y_1	p_{11}	p_{21}	$\ldots$	p_{n1}
y_2	p_{12}	p_{22}	$\ldots$	p_{n2}
$\ldots$	$\ldots$	$\ldots$	$\ldots$	$\ldots$
y_m	p_{1m}	p_{2m}	$\ldots$	p_{nm}

(1.19)

Here p_{ki} is the probability that a ball thrown at the board will get into the slot with coordinates $(x_k; y_i)$. We calculate the probability p_k that a ball gets into a slot with abscissa x_k.

Because a slot with abscissa x_k can have one and only one of the ordinates $y_1, y_2, \ldots, y_m$, we may use (1.17) to derive

$$P_k = p_{k1} + p_{k2} + \cdots + p_{km} \quad (k = 1, 2, \ldots, n). \quad (1.20)$$

EXERCISES

(1) Show that the events A, $(\bar{A}$ and $B)$, $(\overline{A \text{ or } B})$ form a partition.

(2) If the occurrence of an event A implies the occurrence of some other event B, we write $A \subset B$. Show that, in this case, $\mathbf{P}\{A\} \leq \mathbf{P}\{B\}$.

(3) Three cards are drawn from a pack of 52. Calculate the probability that among the cards drawn there will be at least one ace.

(4) An urn contains 5 white and 20 black balls. Balls are drawn from the urn until a white ball appears for the first time. Calculate the probability that this occurs at the third drawing: i.e. that the first white ball is preceded by two black balls.

(5) A workman serves three machines which operate independently of each other. The probability that, in the course of an hour, a machine will not require attention is equal to 0·9 for the first machine, 0·8 for the second machine, and 0·7 for the third machine. Calculate the probability that at least one of the machines will not require service during the course of an hour.

(6) A game between A and B takes place according to the following rules: A makes the first move and will win with probability 0·3; if A does not win, B makes a move and wins with probability 0·5; if B does not win, A makes a second move and wins with probability 0·4; otherwise B wins. If all moves are independent, determine the probabilities of winning for A and B.

(7) Show that the probability of obtaining at least one ace in four independent throws of one die is greater than the probability of obtaining at least one double ace in twenty-four throws of two dice. (This is known as de Méré's problem. It is part of the folk-lore of the subject that probability theory started from de Méré's practical interest in these events.)

(8) Using (1.18), prove that if the events A and B are independent then the events $\bar{A}$ and B are independent.

(9) Show that if $\mathbf{P}\{A|B\} \geqq \mathbf{P}\{A\}$, then also $\mathbf{P}\{B|A\} \geqq \mathbf{P}\{B\}$.

(10) Show that if $\mathbf{P}\{B|A\} = \mathbf{P}\{B|\bar{A}\}$, then (1.8) holds.

(11) Two machines produce identical components: the probability of producing a defective component is 0·03 for the first machine and 0·02 for the second machine. Moreover, the first machine produces per hour twice as many components as the second machine. Calculate the probability that a component chosen at random from the total output of the two machines will not be defective.

(12) A problem concerning four liars. A man, A, receives a piece of information in the form of a signal, "Yes" or "No", which he transmits to a second man, B. B passes on the information to C, and C passes it on to D, who announces it to the world. Each of the four men speaks the truth only once in three times. If D announces the correct information (i.e. the information as A received it), what is the probability that A told the truth?

CHAPTER II

RANDOM VARIABLES AND PROBABILITY DISTRIBUTIONS

§ 6. DISCRETE RANDOM VARIABLES

In this section we shall study functions whose values depend on chance in the sense that they are functions of the outcome of a trial: for example, the number of points appearing on the upper face of a die, or the number of telephone calls at an exchange during a given time interval. The set of all possible values which a given function can take as a result of a trial will be called the *range of possible values* of the function.

DEFINITION. *A function ξ is called a discrete random variable if its range consists of a finite or infinite sequence of values $x_1, x_2, \ldots, x_k, \ldots$ and if the set of those outcomes at which ξ takes the value x_k is an event $(\xi = x_k)$ with a well-defined probability for each k.*

We will denote by p_k the probability of the event $(\xi = x_k)$, and we will say that p_k is the probability of the value x_k. The probability p_k is a function of x_k. This function is called *the probability distribution of the random variable ξ*. It is usual to write this distribution in the form of a table where we enumerate the possible (different) values of the function ξ with their probabilities:

Value of ξ	x_1	x_2	$\cdots$	x_k	$\cdots$
Probability	p_1	p_2	$\cdots$	p_k	$\cdots$

Such an array is known as the *probability distribution table* of the discrete random variable ξ.

If the random variable ξ can take only a finite number of different values $x_1, x_2, \ldots, x_n$, then the events

$$(\xi = x_1), \quad (\xi = x_2), \quad \ldots, (\xi = x_n)$$

form a partition; consequently the sum of their probabilities must be equal to 1:

$$p_1 + p_2 + \cdots + p_n = 1 \tag{2.1}$$

If the distribution table contains an infinite number of values, the condition (2.1) must be written as follows: the infinite series $p_1 + p_2 + \cdots + p_k + \cdots$ converges and its sum must be equal to 1.

We remark that any sequence of numbers $p_1, p_2, \ldots, p_k, \ldots$ satisfying (2.1)—or its modified form—is a possible probability distribution for the random variable ξ.

EXAMPLE 1. The number of points appearing on the upper face of an unbiased die is a discrete random variable with the following probability distribution table (see § 2):

Number of points	1	2	3	4	5	6
Probability	$\frac{1}{6}$	$\frac{1}{6}$	$\frac{1}{6}$	$\frac{1}{6}$	$\frac{1}{6}$	$\frac{1}{6}$

(2.2)

We remark that for a biased die the range will remain the same but the probabilities will no longer all be equal to $\frac{1}{6}$.

EXAMPLE 2. A hunter with three cartridges fires at a target until he makes a hit (or until he has used all three cartridges). The number of used cartridges will be a random variable (ξ) with three possible values (1, 2, 3). We shall find the probability distribution of this function assuming that the shots are independent and that the probability of a hit is 0·8 for each shot.

The event ($\xi = 1$) denotes a hit with the first shot, and its probability is therefore equal to

$$\mathbf{P}\{\xi = 1\} = 0{\cdot}8.$$

The event ($\xi = 2$) denotes a hit with the second shot (having missed with the first) and its probability is given by

$$\mathbf{P}\{\xi = 2\} = (1 - 0{\cdot}8)\, 0{\cdot}8 = (0{\cdot}2)\,(0{\cdot}8) = 0{\cdot}16.$$

Finally, three shots will be used if the first two shots miss, so that

$$\mathbf{P}\{\xi = 3\} = (0{\cdot}2)\,(0{\cdot}2) = 0{\cdot}04.$$

This last probability could also have been derived using (2.1):

$$\mathbf{P}\{\xi = 3\} = 1 - \mathbf{P}\{\xi = 1\} - \mathbf{P}\{\xi = 2\} = 1 - 0{\cdot}8 - 0{\cdot}16 = 0{\cdot}04.$$

Thus the probability distribution table of the function ξ is

ξ	1	2	3
p	0·8	0·16	0·04

(2.3)

EXAMPLE 3. We suppose now that the hunter of the preceeding example has an unlimited supply of cartridges which he fires at a target until he makes a hit. We assume that the probability of a hit with each shot is equal to p, and that the shots are independent. The number of shots required is in this case a random variable with an infinite probability distribution table:

ξ	1	2	3	...	n	...
p	$(1-p)\,p$	$(1-p)^2\,p$		$(1-p)^{n-1}p$		

(2.4)

The sequence of probabilities is an infinite decreasing geometric progression with ratio $(1 - p)$; it converges and its sum is equal to

$$p + (1-p)\,p + (1-p)^2\,p + \cdots + (1-p)^{n-1}p + \cdots$$

$$= \frac{p}{1-(1-p)} = 1.$$

EXAMPLE 4. In several problems of physics and technology we encounter random variables obeying the Poisson distribution

ξ	0	1	2	...	m	...
	e^{-a}	$a\,e^{-a}$	$\frac{a^2}{2!}\,e^{-a}$		$\frac{a^m}{m!}\,e^{-a}$	

(2.5)

where a is a positive constant characteristic of the function ξ.

The Poisson distribution describes, for example: (a) the number of calls at an automatic telephone exchange during a given time interval; (b) the number of electrons given off by an incandescent cathode during a given time-interval. The significance of the constant a will be discussed later. We note here that the series of probabilities $\sum_{m=0}^{\infty} a^m e^{-a}/m!$ converges and its sum is equal to 1:

$$e^{-a}\left(1 + a + \frac{a^2}{2!} + \cdots + \frac{a^m}{m!} + \cdots\right) = e^{-a}e^{+a} = 1.$$

Linear operations on random variables

Operations are called linear which involve the multiplication of a random variable ξ by a number, or the addition of random variables.

If ξ is a discrete random variable with probability distribution

ξ	x_1	x_2	...
	p_1	p_2	...

the product $C\xi$ (where C is a real number) is a discrete random variable with probability distribution

$C\xi$	Cx_1	Cx_2	...
	p_1	p_2	...

(2.6)

In other words, the multiplication of a discrete random variable by a real number involves only the multiplication of the range of values by this number without changing the probabilities.

For example, if the hunter in Ex. 2 (p. 20) pays two shillings for each cartridge he uses, then the sum he spends (in shillings) will be a random variable with the following probability distribution:

2ξ	2	4	6
p	0·8	0·16	0·04

The probability distribution of the sum of two discrete random variables is somewhat more difficult to determine. Let us suppose that ξ and η are two discrete random variables with probability distributions

ξ	x_1	x_2	$\dots$
	p_1	p_2	$\dots$

η	y_1	y_2	$\dots$
	q_1	q_2	$\dots$

and let us denote by p_{kl} the probability of the intersection of the events $(\xi = x_k)$ and $(\eta = y_l)$. If this intersection occurs, then the sum $\xi + \eta$ takes the value $x_k + y_l$. Now it is clear for example, that the probability of the event $(\xi + \eta = x_1 + y_1)$ can be greater than p_{11} if for some other values of k and l the corresponding sums $x_k + y_l$ are equal to $x_1 + y_1$. Thus *the range of possible values of the sum $\xi + \eta$ is the set of all possible sums of values of ξ and values of η, and the probability of any one of these values of the sum is the sum of the probabilities of all the intersections $((\xi = x_k)$ and $(\eta = y_l))$ in which the given sum is attained.*

In order to construct the probability distribution table of the sum $\xi + \eta$ it is useful in practice to construct first an auxiliary table

$x_1 + y_1$	$x_1 + y_2$	$x_2 + y_1$	$x_2 + y_2$	$\dots$
p_{11}	p_{12}	p_{21}	p_{22}	$\dots$

(2.7)

and then to unite the entries with the same value of $x_k + y_l$, adding the corresponding probabilities p_{kl}. We remark that if all the values $x_k + y_l$ are different, then the table (2.7) will be complete as a distribution table for the sum $\xi + \eta$.

EXAMPLE. Let us perform the experiment of throwing, simultaneously, two unbiased dice, and let ξ be the number appearing on the first die, η the number appearing on the second die, so that $\xi + \eta$ is the sum of the numbers appearing on the two dice.

Both the random variables ξ and η have the same probability distribution table (2.2): we say that ξ and η are *identically distributed.* We will determine the distribution of their sum $\xi + \eta$. Because the numbers appearing on the two dice are essentially independent, the probability of each intersection will be equal to 1/36. We construct, therefore, an auxiliary table on the model of (2.7):

$1+1$	$1+2$	$2+1$	$1+3$	$2+2$	$3+1$	$\ldots$	$6+6$
$\frac{1}{36}$	$\frac{1}{36}$	$\frac{1}{36}$	$\frac{1}{36}$	$\frac{1}{36}$	$\frac{1}{36}$	$\ldots$	$\frac{1}{36}$

Uniting the entries with the same value in the first line—the line of possible values—we obtain the probability distribution table of the random variable $\xi + \eta$:

2	3	4	5	6	7	8	9	10	11	12
$\frac{1}{36}$	$\frac{2}{36}$	$\frac{3}{36}$	$\frac{4}{36}$	$\frac{5}{36}$	$\frac{6}{36}$	$\frac{5}{36}$	$\frac{4}{36}$	$\frac{3}{36}$	$\frac{2}{36}$	$\frac{1}{36}$

We call the reader's attention to the following peculiarity of the addition of random variables. Comparing the probability distribution table of the sum $\xi + \eta$ with the probability distribution table of 2ξ,

2	4	6	8	10	12
$\frac{1}{6}$	$\frac{1}{6}$	$\frac{1}{6}$	$\frac{1}{6}$	$\frac{1}{6}$	$\frac{1}{6}$

shows that, in general, the addition of identically distributed random variables is not the same as multiplying one of them by a whole number.

It is easily verified that the operations of addition and multiplication by a real number preserve for random variables the usual properties of addition and multiplication for real numbers:

$$\xi + \eta = \eta + \xi; \quad (\xi + \eta) + \zeta = \xi + (\eta + \zeta)$$

$$C(\xi + \eta) = C\xi + C\eta.$$

Independence of random variables

The discrete random variables ξ and η are called *independent* if the events $(\xi = x_k)$ and $(x = y_l)$ are independent for all k and l, i.e. if the probability of the intersection of these events can be written in the form

$$p_{kl} = p_k q_l \quad (k = 1, 2, \ldots; \ l = 1, 2, \ldots). \tag{2.8}$$

For example, in throwing two dice the numbers appearing on the two dice are independent random variables; this simplifies the argument in the above example.

Random variables $\xi_1, \xi_2, \ldots, \xi_n$ are called *mutually independent* if all the events $(\xi_1 = x_{k_1}^{(1)})$, $(\xi_2 = x_{k_2}^{(2)}), \ldots, (\xi_n = x_{k_n}^{(n)})$ are mutually independent (where $x_1^{(i)}, x_2^{(i)}, \ldots, x_{k_i}^{(i)}, \ldots$ denotes the range of the function ξ_i).

If the random variables $\xi_1, \xi_2, \ldots, \xi_n$ are mutually independent, then it is easy to find the probability distribution of an arbitrary linear combination $C_1\xi_1 + C_2\xi_2 + \cdots + C_n\xi_n$ ($C_1, C_2, \ldots, C_n$ constants) in terms of the probability distributions of the random variables $\xi_1, \xi_2, \ldots, \xi_n$. We make use of this observation in calculations where it is often convenient to represent a random variable in which we are interested by a linear combination of independent random variables.

§ 7. THE BINOMIAL DISTRIBUTION

In this section we shall discuss the relative frequency, ω_n, of an event A in n repeated trials. We shall suppose that the occurrence of A at each trial is independent of its occurrence at the other trials, and also that the probability of A is the same for all trials. We shall denote this number by p.†

† We can represent these repeated trials in the following way. Place a number of identical balls in an urn and then mark some fraction of them "A" (colour them white, for example); the proportion of marked balls must be equal to p, so that the probability of drawing a marked ball is equal to the probability of the event A. Draw one ball at random from the urn and record whether or not it is marked. Return the ball to the urn, carefully mixing the balls, and then draw again. This process is repeated until we have n readings. Such a sequence of trials is called *a sequence of independent Bernoulli trials*, or *random sampling with replacement* (to be distinguished from the scheme in which the ball is not replaced; e.g. Example 1, p. 16).

In n trials the event A can occur $0, 1, 2, \ldots, n$ times, so that ω_n will be a discrete random variable with range $0, 1/n, 2/n, \ldots, 1$. We will evaluate its probability distribution. To this end we represent the random variable ω_n as a linear combination of simpler functions. We introduce the *indicator random variable* λ_k, which is the number of times the event A occurs at the kth trial. The random variable λ_k can take only two values: 1, if the event A occurs at the kth trial, and 0, if the event A does not occur at the kth trial. Because the probability of the event A is equal to p for all trials, the random variables $\lambda_1, \lambda_2, \ldots, \lambda_n$ will all have the same probability distribution table

$$\lambda_k \quad \begin{array}{c|c} 1 & 0 \\ \hline p & q \end{array} \tag{2.9}$$

where $q = 1 - p$. It follows directly from our assumptions that the random variables $\lambda_1, \lambda_2, \ldots, \lambda_n$ are mutually independent.

Let us now consider the sum of the indicator random variables

$$\mu_n = \lambda_1 + \lambda_2 + \cdots + \lambda_n. \tag{2.10}$$

This sum consists entirely of zeros and ones, and the number of ones is equal to the number of times the event A occurs in the n trials: consequently the random variable μ_n is equal to the number of repetitions of the event A in n trials and the ratio of μ_n to the number of trials is the relative frequency ω_n:

$$\omega_n = \frac{\mu_n}{n} = \frac{1}{n}(\lambda_1 + \lambda_2 + \cdots + \lambda_n). \tag{2.11}$$

Having established ω_n in the form of a linear combination of mutually independent random variables with known distributions, we can find the probability distribution of ω_n using the ideas of the preceding section. We will add the functions $\lambda_1, \lambda_2, \ldots, \lambda_n$ one at a time. First of all, using (2.7) and (2.8) we have

$$\lambda_1 \quad \begin{array}{c|c} 1 & 0 \\ \hline p & q \end{array} + \lambda_2 \quad \begin{array}{c|c} 1 & 0 \\ \hline p & q \end{array} = \mu_2 \quad \begin{array}{c|c|c} 2 & 1 & 0 \\ \hline pp & pq + qp & qq \end{array},$$

or

μ_2	2	1	0
	p^2	$2pq$	q^2

.

Continuing in the same way we find

$\mu_2 + \lambda_3 = \mu_3$	3	2	1	0
	p^2p	$p^2q + 2pqp$	$2pqq + q^2p$	q^2q

or

μ_3	3	2	1	0
	p^3	$3p^2q$	$3pq^2$	q^3

.

We see that the probabilities in the distribution tables of the functions μ_2 and μ_3 are precisely the corresponding terms in the binomial expansions

$$(p + q)^2 = p^2 + 2pq + q^2,$$

$$(p + q)^3 = p^3 + 3p^2q + 3pq^2 + q^3$$

(whence, by the way, it is immediately evident that the sums of the probabilities in these tables are equal to one since $p + q = 1$).

Using the method of induction we shall prove the following general assertion: *the probability that μ_n takes some value m is equal to the term containing p^m in the expansion of $(p + q)^n$ in powers of p*:

$$\mathbf{P}\{\mu_n = m\} = \binom{n}{m} p^m q^{n-m} = \frac{n!}{m!(n-m)!} p^m q^{n-m}. \quad (2.12)$$

Proof: The validity of the expression (2.12) for $n = 2$ (and also for $n = 3$) has already been established. We show that if (2.12) is valid for some number n, then its validity for $n + 1$ follows. In fact, the random variable $\mu_{n+1} = \mu_n + \lambda_{n+1}$ can take the value m in only two ways: either $\mu_n = m$ and $\lambda_{n+1} = 0$, or $\mu_n = m - 1$ and $\lambda_{n+1} = 1$. Therefore, since these possibilities are mutually exclusive, and using the mutual independence of the $\lambda_1, \lambda_2, \ldots, \lambda_{n+1}$, we have

$$\mathbf{P}\{\mu_{n+1} = m\} = \mathbf{P}\{\mu_n = m\}\,\mathbf{P}\{\lambda_{n+1} = 0\} + \mathbf{P}\{\mu_n = m - 1\}\,\mathbf{P}\{\lambda_{n+1} = 1\}$$

$$= \frac{n!}{m!\,(n-m)!} p^m q^{n-m} q + \frac{n!}{(m-1)!\,(n-m+1)!} p^{m-1} q^{n-m+1} p$$

so that

$$\mathbf{P}\{\mu_{n+1} = m\} = \frac{(n+1)!}{m!\,(n-m+1)!} p^m q^{n-m+1},$$

which was to be proved.

We can derive (2.12) without having to represent μ_n in the form (2.10). Using the assumption of independence, the probability that the event A occurs in the first m trials and does not occur in the remaining $n - m$ trials is seen to be

$$p^m q^{n-m}. \tag{2.13}$$

However, $\mathbf{P}\{\mu_n = m\}$ does not depend on which of the n trials the m events A actually occur, and in n consecutive trials there are $\binom{n}{m}$ different ways in which it is possible for the event A to occur m times. Since these $\binom{n}{m}$ ways are mutually exclusive we add the probabilities to obtain the desired probability of the event $(\mu_n = m)$, which is therefore equal to the probability (2.13) multiplied by $\binom{n}{m}$, i.e. we have once again obtained the expression (2.12).

The expression (2.12) gives the probability that in n repeated trials the event A will occur precisely m times. Thus we obtain the following probability distribution tables for the random variables μ_n and ω_n:

μ_n	n	$n-1$	$\ldots$	m	$\ldots$	1	0
	p^n	$np^{n-1}q$		$\binom{n}{m} p^m q^{n-m}$		npq^{n-1}	q^n

(2.14)

ω_n	1	$\frac{n-1}{n}$	$\ldots$	$\frac{m}{n}$	$\ldots$	$1/n$	0
	p^n	$np^{n-1}q$		$\binom{n}{m} p^m q^{n-m}$		npq^{n-1}	q^n

(2.15)

The probability distribution given by table (2.14) is known as the *binomial distribution.*

EXAMPLE. From a large batch of manufactured articles, 10 are selected for test. It is known that the proportion of defective items in the whole batch is 25%; we wish to find the probability that more than 5 of the selected articles will be defective.

The selection of each article will constitute a trial: if the article is found to be defective we shall say that the event A has occurred. The probability, p, of the event A is evidently equal to the proportion of defective items in the whole batch, i.e. $p = 0{\cdot}25$. The number of defective items among the ten selected items is the random variable μ_{10}, i.e. the frequency of occurrence of the event A in ten trials. The problem reduces to the calculation of the probability that $\mu_{10} > 5$; using the additivity property we find

$$\mathbf{P}\{\mu_{10} > 5\} = \mathbf{P}\{\mu_{10} = 6\} + \mathbf{P}\{\mu_{10} = 7\} + \mathbf{P}\{\mu_{10} = 8\} + \mathbf{P}\{\mu_{10} = 9\} + \mathbf{P}\{\mu_{10} = 10\}.$$

The following distribution table has been calculated using the binomial term (2.12), with $p = 0{\cdot}25$, $q = 0{\cdot}75$, and $n = 10$ (the probabilities have been rounded off to 0·0001):

The number of defective items, m	Probability $\mathbf{P}\{\mu_{10} = m\}$
0	0·0563
1	0·1877
2	0·2816
3	0·2503
4	0·1460
5	0·0584
6	0·0162
7	0·0031
8	0·0004
9	0·0000
10	0·0000

From this it is clear that $\mathbf{P}\{\mu_{10} > 5\} \approx 0{\cdot}020$. This probability is sufficiently small that if we should find six (or more) defectives among the ten items selected, then we could doubt whether the proportion of defective items in the whole batch is really as low as 25%.

Note: Our solution of this problem depends upon the random variable μ_{10} having a binomial distribution, and this is assured only when the articles are selected for testing according to the scheme of random sampling with replacement (see footnote, p. 25). To clarify the difference between sampling with replacement and sampling without replacement, let us consider the following example. We suppose that there are $N = 100$ balls in an urn of which $M = 25$ are white (so that $p = 0{\cdot}25$) and we draw 2 balls from the urn one after the other. We compare the probability distributions of the number of white balls drawn for sampling with replacement and sampling without replacement. In both cases the number of white balls drawn is the sum of two indicator random variables λ_1 and λ_2, where λ_k is the number of white balls obtained at the kth draw; λ_1 and λ_2 have the same probability distribution (see Ex. 1, p. 16):

λ_1	1	0
	$\frac{25}{100}$	$\frac{75}{100}$

λ_2	1	0
	$\frac{25}{100}$	$\frac{75}{100}$

However, when the sampling is with replacement λ_1 and λ_2 will be independent, whereas when the sampling is without replacement they will be dependent. Consequently the probability distribution of their sum is given by the tables

μ_2	2	1	0
	$\left(\frac{25}{100}\right)^2$	$2 \times \frac{25}{100} \times \frac{75}{100}$	$\left(\frac{75}{100}\right)^2$

when the sampling is with replacement, and

μ_2	2	1	0
	$\frac{25}{100} \times \frac{24}{99}$	$\frac{25}{100} \times \frac{75}{99} + \frac{75}{100} \times \frac{25}{99}$	$\frac{75}{100} \times \frac{74}{99}$

when the sampling is without replacement.

Comparing these distribution tables we see that the corresponding probabilities differ little from each other. Clearly this difference would be even smaller if we had taken, for example, $N = 1{,}000$, $M = 250$. Similar reasoning leads to the conclusion that, the larger the numbers N and M in relation to the sample size, the more accurate will be the solution based on the binomial distribution.

§ 8. CONTINUOUS RANDOM VARIABLES

Discrete random variables do not form the only type of random variable: in probability theory we often encounter random variables whose ranges of possible values are entire intervals. For example, in the Introduction we discussed the problem of measuring the deviation from a central value of the dimensions of machined components. Such random variables are called *continuous.*†

The events of interest to us are of the form $(x_1 < \xi < x_2)$, and the probability distribution of ξ must enable us to find the probability of this event for any interval (x_1, x_2). We shall denote this probability by $\mathbf{P}\{x_1 < \xi < x_2\}$.

EXAMPLE. The uniform distribution. In the simplest case the range of a random variable is a finite interval (α_1, α_2), and if (x_1, x_2) is any interval lying inside (α_1, α_2), then the probability $\mathbf{P}\{x_1 < \xi < x_2\}$ is proportional to the length of this interval:

$$\mathbf{P}\{x_1 < \xi < x_2\} = \lambda(x_2 - x_1) \quad (\alpha_1 \leqq x_1 < x_2 \leqq \alpha_2). \tag{2.16}$$

We must choose the coefficient λ so that the second fundamental property of probabilities is satisfied;‡ since the range of all possible values of ξ is the interval (α_1, α_2), the event $(\alpha_1 < \xi < \alpha_2)$ is the certain event and so must have probability equal to 1:

$$\mathbf{P}\{\alpha_1 < \xi < \alpha_2\} = \lambda(\alpha_2 - \alpha_1) = 1$$

thence

$$\lambda = \frac{1}{\alpha_2 - \alpha_1}.$$

If the probability distribution of a random variable is given by (2.16), we say that ξ is *uniformly distributed on the interval* (α_1, α_2), or that the random variable ξ *has the uniform probability distribution.*

† We shall give precise conditions later.

‡ The first fundamental property requires that the coefficient λ be positive, and the third expresses the fact that if we unite two non-overlapping intervals then we add their lengths.

The probability density function

If the random variable ξ is uniformly distributed on the interval (α_1, α_2), then for any two points x_1, x_2 of this interval the ratio

$$\frac{\mathbf{P}\{x_1 < \xi < x_2\}}{x_2 - x_1} \tag{2.17}$$

of the probability $\mathbf{P}\{x_1 < \xi < x_2\}$ to the length of the interval (x_1, x_2) is a constant equal to $\lambda = 1/(\alpha_2 - \alpha_1)$. This ratio is called the probability density for the uniformly distributed random variable ξ.

For an arbitrary continuous random variable ξ the ratio (2.17) will no longer be a constant. Using the analogy of mechanics (where mass distributions are studied) we introduce the idea of a probability density at a point.

The limit of the ratio

$$\lim_{\Delta x \to 0} \frac{\mathbf{P}\{x < \xi < x + \Delta x\}}{\Delta x} = \varphi(x) \tag{2.18}$$

is called the probability density of the random variable ξ at the point x.

We shall consider only those random variables for which this limit exists at each point x. For such a random variable the probability of ξ taking a value in the interval $(x, x + dx)$ can be expressed approximately by a principal term proportional to dx:

$$\mathbf{P}\{x < \xi < x + dx\} \approx \varphi(x)\, dx$$

(ignoring terms of a higher order in dx).

This principal term is known as the *probability differential*, and is denoted by $d\mathbf{P}_x$:

$$d\mathbf{P}_x = \varphi(x)\, dx. \tag{2.19}$$

Knowing the probability differential we can derive the probability of the event $(x_1 < \xi < x_2)$ by integrating:

$$\mathbf{P}\{x_1 < \xi < x_2\} = \int_{x_2}^{x_1} \varphi(x)\, dx. \tag{2.20}$$

Thus, to be able to determine completely the probability distribution of a continuous random variable, it is sufficient to know its probability density function (i.e. the function $\varphi(x)$).† In all calculations with continuous random variables the probability differential $\varphi(x)\,dx$ plays the role played by the probabilities p_k in our work with discrete random variables. In many formulae it will be sufficient to replace p_k by $\varphi(x)\,dx$ and to replace the sum by an integral in order to turn expressions for discrete random variables into expressions for continuous random variables.

Remark. We emphasise that for a continuous random variable ξ the events of real significance are those of the form $(x_1 < \xi < x_2)$, and not $(\xi = x)$. Since the probability of ξ taking a value in some small interval is approximately proportional to the length of the interval it follows that *the probability of ξ taking the value of a particular point is equal to zero.* In other words, for a continuous random variable ξ an event of the form $(\xi = x)$ has probability zero (although such an event cannot be considered as impossible). In practice the above statement does not lead to a misunderstanding because we can only measure the value of a physical variable with limited accuracy (an absolutely precise value of a physical variable is only a mathematical abstraction).

Fundamental properties of the probability density function

(a) The density function $\varphi(x)$ is non-negative for all x. This follows immediately from (2.18) where $\Delta x > 0$ and

$$\mathbf{P}\{x < \xi < x + \Delta x\} \geqq 0.$$

(b) The integral of the density function taken over the whole range of the random variable ξ is equal to 1. This follows from the fact that as a result of a trial ξ takes some value in its range. so that the integral is the probability of the certain event. We write the condition in the form

$$\int_{\alpha_1}^{\alpha_2} \varphi(x)\,dx = 1 \quad \text{or} \quad \int_{-\infty}^{\infty} \varphi(x)\,dx = 1$$

according as the range is the finite interval (α_1, α_2) or the entire real axis. We unite both these expressions by writing

$$\int \varphi(x)\,dx = 1 \tag{2.21}$$

without indicating the limits of integration, on the understanding that the integral is taken over the whole range of possible values of the random variable ξ.

† Strictly speaking, a continuous random variable ξ is characterised by this, that the probability of it taking a value in an arbitrary interval (x_1, x_2) can be represented in the form of an integral (2.20) for some function $\varphi(x)$.

We remark that any non-negative function $\varphi(x)$ satisfying (2.21) can be the probability density function for some random variable ξ.

Probability curves

The graph of the function $y = \varphi(x)$ will be called *the probability curve* of a random variable having $\varphi(x)$ as probability density function. We can use a probability curve to calculate

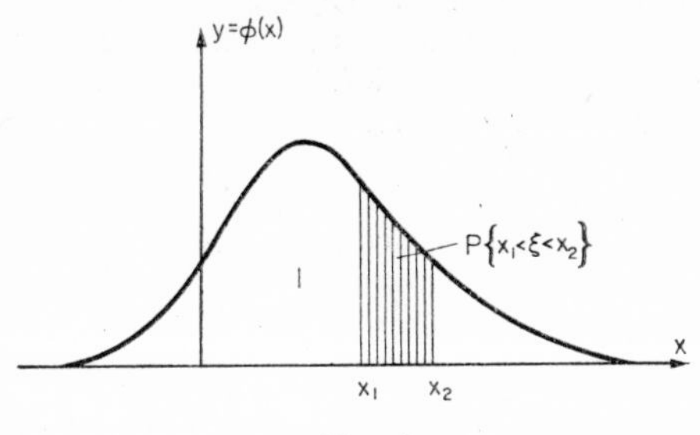

FIG. 3

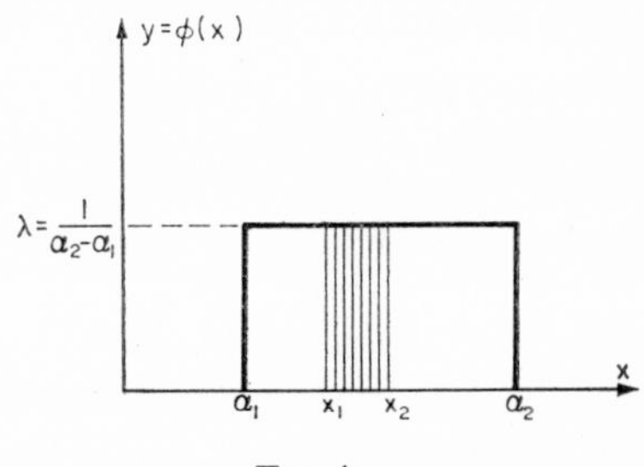

FIG. 4

probabilities graphically, since the probability $\mathbf{P}\{x_1 < \xi < x_2\}$ and the hatched area of the trapezium in Fig. 3 are both expressed by the integral (2.20). In general we will not draw the graph so that the area under the curve is actually equal to one. For graphical calculations therefore *the probability* $\mathbf{P}\{x_1 < \xi < x_2\}$ *is equal to the ratio of the hatched area to the area under the entire curve.* Since a probability is a dimensionless number, it follows that the dimension of $\varphi(x)$ is equal to 1/dimension of x.

As an example we give in Fig. 4 the curve of the uniform probability distribution on the interval (α_1, α_2).

EXAMPLES. *Continuous Probability Distributions.*

(1) The standard Normal Distribution. We shall say that the random variable ξ_0 has *the standard normal distribution or that* ξ_0 is *a standard normal variable* if its range is all real numbers from $-\infty$ to $+\infty$ and its density is defined to be

$$\varphi_0(x) = C\, e^{-x^2/2}, \quad \text{where} \quad C = \frac{1}{\sqrt{(2\pi)}}. \tag{2.22}$$

The value of the coefficient C is chosen to satisfy (2.21). The probability curve is given in Fig. 5.† It is symmetric with respect

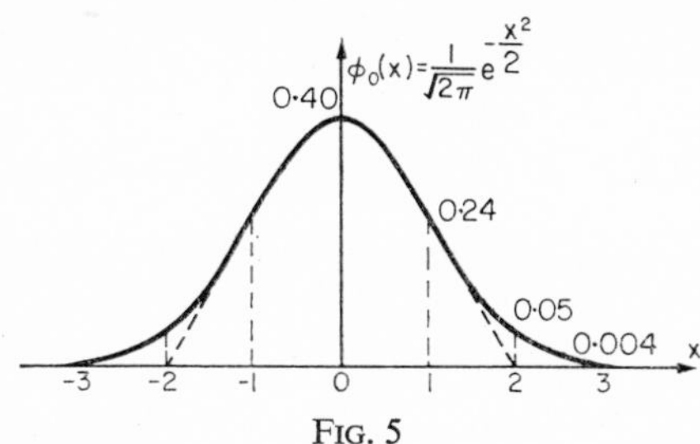

FIG. 5

to the y axis, it achieves its maximum value of $1/\sqrt{(2\pi)} \approx 0{\cdot}4$ at $x = 0$, and it has two points of inflexion at $x = \pm 1$. As $x \to \pm\infty$ the probability curve approaches the x axis asymptotically (and moreover approaches very fast; for example, $\varphi_0(3) = 0{\cdot}0044$, $\varphi_0(4) = 0{\cdot}00013$).

The normal distribution plays an important role in many applications of probability theory, in particular to the theory of observations (see Chapters V and VI). Because the integral of the density $\varphi_0(x)$ cannot be expressed finitely in terms of elementary functions, the calculation of probabilities is carried out using the detailed tables of the function

$$\Phi(t) = \frac{2}{\sqrt{(2\pi)}} \int_0^t e^{-x^2/2}\, dx, \tag{2.23}$$

known as *the normal probability integral.* A short table of the normal probability integral is given in the Appendix (Table I). The function $\Phi(t)$ is an odd function, since

$$\Phi(-t) = \frac{2}{\sqrt{(2\pi)}} \int_0^{-t} e^{-x^2/2}\, dx = -\Phi(t);$$

† See also Table III.

consequently the tables only give the values of $\Phi(t)$ for positive values of t. As t goes from 0 to ∞ the function $\Phi(t)$ increases from 0 to 1, and moreover it increases very fast: thus $\Phi(3) = 0{\cdot}9973$, $\Phi(4) = 0{\cdot}999937$. The graph of $\Phi(t)$ is shown in Fig. 6.

With the help of the function $\Phi(t)$ it is possible to calculate the probability that the random variable ξ_0 takes a value in the interval (x_1, x_2). We have

$$\mathbf{P}\{x_1 < \xi_0 < x_2\}$$

$$= \int_{x_1}^{x_2} \varphi_0(x)\, dx = \frac{1}{\sqrt{(2\pi)}} \int_0^{x_2} e^{-x^2/2}\, dx - \frac{1}{\sqrt{(2\pi)}} \int_0^{x_1} e^{-x^2/2}\, dx,$$

so that

$$\mathbf{P}\{x_1 < \xi_0 < x_2\} = \tfrac{1}{2}\Phi(x_2) - \tfrac{1}{2}\Phi(x_1). \tag{2.24}$$

In particular, for the symmetric interval $(-t, t)$ we derive

$$\mathbf{P}\{-t < \xi_0 < t\} = \tfrac{1}{2}\Phi(t) - \tfrac{1}{2}\Phi(-t) = \Phi(t). \tag{2.25}$$

Thus, for $t > 0$, the function $\Phi(t)$ gives the probability that the standard normal variable ξ_0 takes a value in the interval $(-t, t)$.

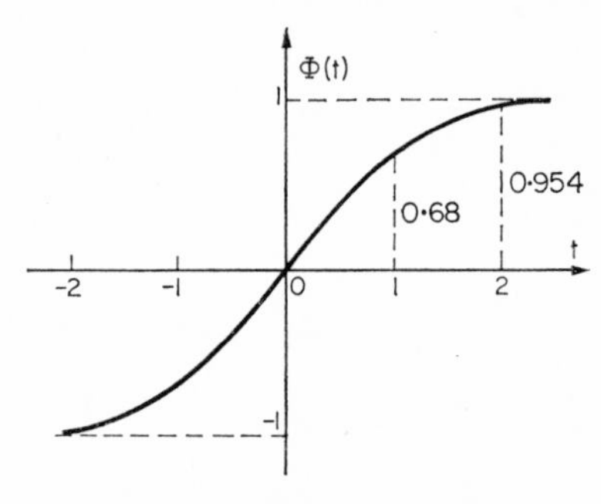

FIG. 6

(2) The general Normal Distribution. This name is given to the distribution with density

$$\varphi(x) = \frac{1}{\sigma\sqrt{(2\pi)}} \exp\left(-\frac{(x-a)^2}{2\sigma^2}\right), \tag{2.26}$$

where $\sigma > 0$. For $a = 0$ and $\sigma = 1$ the density $\varphi(x)$ reduces to $\varphi_0(x)$, the density of the standard normal distribution. The curves of the general normal distribution are given in Fig. 7 for various values of σ (and $a = 0$). They differ from the standard normal distribution (2.22) only in a change of scale along the axis. With increasing σ the probability curves become more gently sloping. A change in a displaces the probability curve along the x- axis (see Fig. 8); this curve is symmetric about the point $x = a$. In the next section we shall consider the question of calculating probabilities with the general normal distribution (see p. 42).

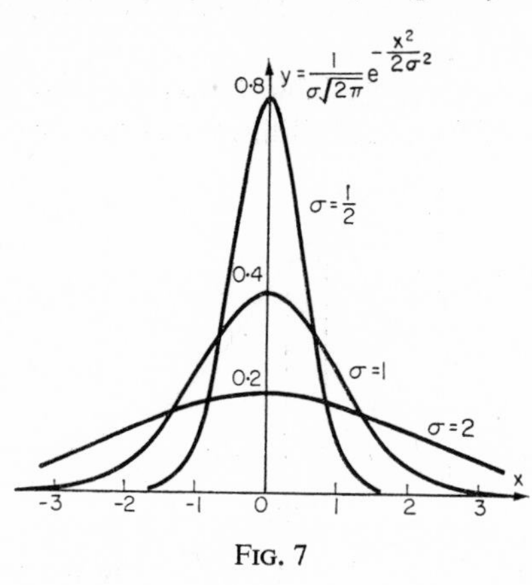

FIG. 7

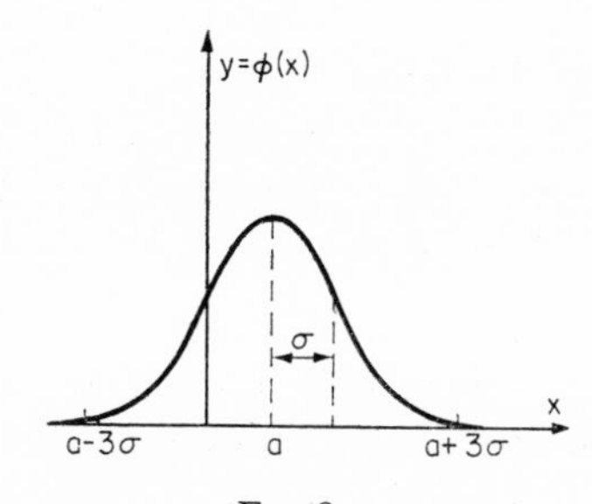

FIG. 8

(3) As an example of an unsymmetric probability distribution we introduce

$$\varphi(x) = \begin{cases} 0 & \text{for } x \leq 0 \\ C_1 x^{\alpha-1} e^{-\beta x} & \text{for } x > 0 \quad (\alpha > 0, \beta > 0). \end{cases} \tag{2.27}$$

The coefficient C_1 is chosen so that (2.21) is satisfied:

$$C_1 = \frac{\beta^\alpha}{\Gamma(\alpha)}, \quad \text{where } \Gamma(\alpha) = \int_0^\infty x^{\alpha-1} e^{-x}\, dx$$

is Euler's Gamma function. The distribution (2.27) belongs to the so-called Pearson distributions; it occurs, for example, in problems arising in water-power engineering. Figure 9 shows the probability curve for (2.27) with $\alpha = \beta = 4$. The calculation of probabilities is also carried out for these distributions with the aid of special tables.†

The cumulative distribution function

The cumulative distribution function, $F(x)$, is the probability that the random variable ξ takes a value less than some number x; i.e. $F(x) = \mathbf{P}\{\xi < x\}$.

For a discrete random variable the cumulative distribution is equal to the sum of the probabilities for all values of x_k which are

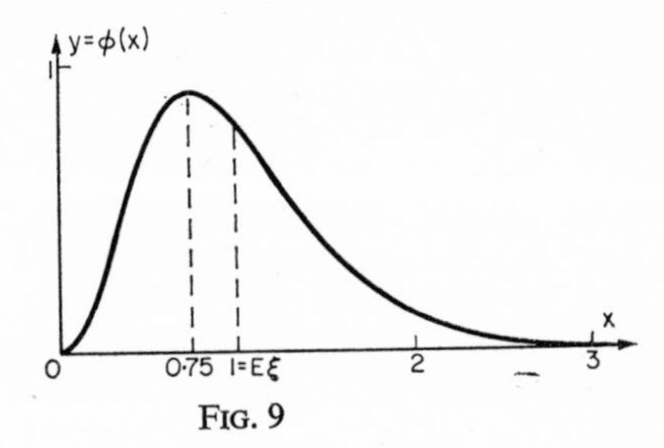

FIG. 9

less than x; $F(x) = \sum_{x_k<x} p_k$. For example, from the probability distribution table (2.3) we obtain

$$F(x) = \begin{cases} 0 & \text{for} \quad x \leqq 1 \\ 0{\cdot}8 & \text{for} \quad 1 < x \leqq 2 \\ 0{\cdot}96 & \text{for} \quad 2 < x \leqq 3 \\ 1 & \text{for} \quad 3 < x. \end{cases}$$

For a continuous random variable the expression (2.20) indicates that the cumulative distribution function is given by

$$F(x) = \int_{-\infty}^{x} \varphi(t)\, dt.$$

For example, in the case of the standard normal distribution the cumulative distribution is expressed in terms of the normal integral:

$$F(x) = \int_{-\infty}^{x} \varphi_0(t)\, dt = \tfrac{1}{2}\Phi(x) - \tfrac{1}{2}\Phi(-\infty) = \tfrac{1}{2}\Phi(x) + \tfrac{1}{2}.$$

† *Tables of the Incomplete Γ-function*, ed. K. Pearson. H.M.S.O., 1922.

From the definition of $F(x)$ and the fundamental properties of probabilities we deduce that the cumulative distribution function is an increasing function lying between 0 and 1. Its graph is called *the cumulative probability curve* (Fig. 10). In Fig. 11 we give as an example the cumulative probability curve of the uniform distribution on the interval (α_1, α_2). From the additivity property we have for $x_1 < x_2$,

$$\mathbf{P}\{\xi < x_2\} = \mathbf{P}\{\xi < x_1\} + \mathbf{P}\{x_1 \leqq \xi < x_2\},$$

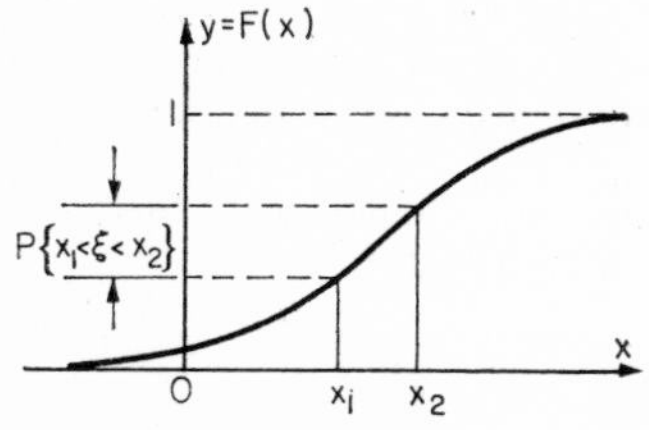

FIG. 10

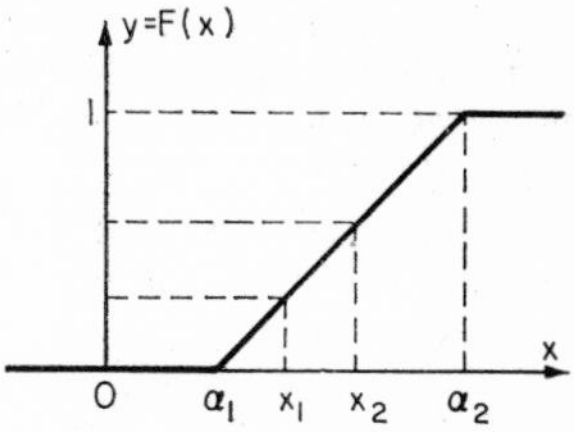

FIG. 11

so that the probability of ξ taking a value in the interval (x_1, x_2) is equal to the increment of the cumulative distribution function:

$$\mathbf{P}\{x_1 \leqq \xi < x_2\} = F(x_2) - F(x_1).$$

This enables us to obtain probabilities graphically from the cumulative distribution curve (Fig. 10), if the scale is properly chosen so that $F(+\infty) = 1$.

The cumulative distribution function is suitable for discussing problems involving both discrete and continuous random variables (and also random variables of a more complicated nature). However, its application generally requires special mathematical apparatus (the Stieltjes integral) which lies outside the scope of this book.

§ 9. FUNCTIONS OF RANDOM VARIABLES

Let $f(x)$ be a single valued function defined on the range of the random variable ξ. Then the function $f(\xi)$ of the random variable ξ is itself a random variable η which takes the value $y = f(x)$ when the random variable ξ takes the value x. For example, if the random variable ξ is the diameter of a cylinder being turned on a lathe, then the area of cross-section of the cylinder is the random variable $\eta = \pi\xi^2/4$.

We wish to establish the connection between the probability distributions of the random variables ξ and $\eta = f(\xi)$. We begin with a function of the discrete random variable

ξ	x_1	x_2	$\dots$
	p_1	p_2	$\dots$

If as the result of a trial the random variable ξ takes the value x_k, then the random variable $\eta = f(\xi)$ takes the value $f(x_k)$. But the probability of the event $(\eta = f(x_1))$, say, can be bigger than the probability p_1 of the event $(\xi = x_1)$ if there occur among the values $f(x_k)$ others which are equal to $f(x_1)$. Thus using the additivity property we see that the probability of the event $(\eta = f(x_1))$ is equal to the sum of all those probabilities p_k for which the numbers $f(x_k)$ are equal to $f(x_1)$.

In order to construct the probability distribution table for the function $f(\xi)$ it is useful in practice to construct first an auxiliary table

$f(x_1)$	$f(x_2)$	$\dots$
p_1	p_2	$\dots$

(2.28)

and then to unite the equal values of $f(x_k)$, at the same time adding the corresponding probabilities. We remark that if all the values $f(x_k)$ are different, then table (2.28) will be complete as a distribution table for the function $f(\xi)$ (cf. (2.6)).

EXAMPLE 1. Consider the powers λ^n $(n = 1, 2, 3, \dots)$ of an indicator random variable λ with probability distribution (2.9). All the powers will have the same distribution as λ because $1^n = 1$, $0^n = 0$.

EXAMPLE 2. Consider the function $\sin[(\pi/2)\xi]$ of the random variable

ξ	1	2	3	$\dots$	n	$\dots$
	$\frac{1}{2}$	$\frac{1}{2^2}$	$\frac{1}{2^3}$	$\dots$	$\frac{1}{2^n}$	$\dots$

Because

$$\sin[(\pi/2)n] = \begin{cases} 0 & \text{for even } n, \\ 1 & \text{for } n = 4k+1, \\ -1 & \text{for } n = 4k+3, \end{cases}$$

the distribution table for $\sin[(\pi/2)\xi]$ will be

$\sin[(\pi/2)\xi]$	0	1	-1
	p_0	p_1	p_{-1}

where

$$p_0 = \frac{1}{2^2} + \frac{1}{2^4} + \frac{1}{2^6} + \cdots = \frac{1}{4(1-\frac{1}{4})} = \frac{1}{3};$$

$$p_1 = \frac{1}{2} + \frac{1}{2^5} + \frac{1}{2^9} + \cdots = \frac{1}{2(1-\frac{1}{16})} = \frac{8}{15};$$

$$p_{-1} = \frac{1}{2^3} + \frac{1}{2^7} + \frac{1}{2^{11}} + \cdots = \frac{1}{8(1-\frac{1}{16})} = \frac{2}{15}.$$

EXAMPLE 3. Consider the square of the random variable

ξ	$-b$	b
	p	$1-p$

. Because $(-b)^2 = b^2$ we obtain

ξ^2	b^2
	1

.

From this table we see that ξ^2 takes the single value b^2 with probability 1 so that it can no longer be regarded as a genuine random variable.

Functions of a continuous random variable

We will consider a function $f(x)$ which is continuous and possesses a first derivative at all points x of the range of the random variable ξ. The problem in this case is to establish the relationship between the probability density functions $\varphi(x)$ and $\psi(x)$ of the random variables ξ and $\eta = f(\xi)$ respectively. We suppose first of all that $f(x)$ is a *strictly increasing function*. Thus each interval (x_1, x_2) is in a one-to-one relationship with a corresponding interval (y_1, y_2)

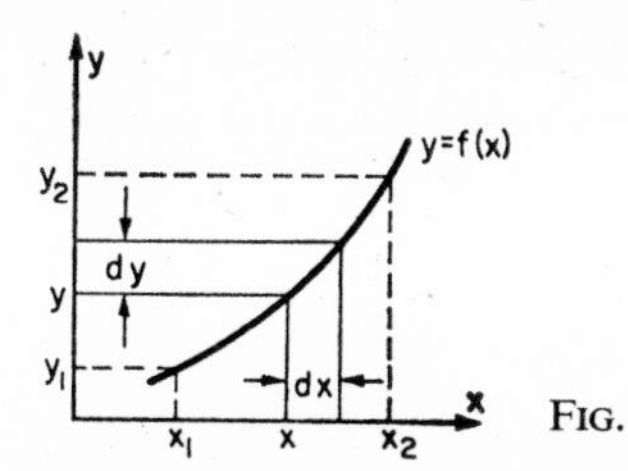

FIG. 12

(see Fig. 12), and therefore the probabilities of the random variables ξ and η taking values in such corresponding intervals must be equal. If the intervals $(x, x + dx)$ and $(y, y + dy)$ are related in this way, this means that the probability differentials are equal:

$$\varphi(x)\, dx = \psi(y)\, dy, \tag{2.29}$$

from which we deduce that

$$\psi(y) = \varphi(x)\frac{dx}{dy} = \varphi[g(y)]\, g'(y), \tag{2.30}$$

where $x = g(y)$ is the inverse of the function $y = f(x)$.

If the function $y = f(x)$ is *strictly decreasing* then a positive value of dx corresponds to a negative value of dy. It is therefore necessary to replace dy by $-dy = |dy|$ in (2.29) so that (2.30) becomes

$$\psi(y) = \varphi(x) \left|\frac{dx}{dy}\right| = \varphi[g(y)]\,|g'(y)|. \tag{2.31}$$

EXAMPLE. *The linear function.* For the linear function $\eta = a + b\xi$ we have:

$$y = f(x) = a + bx; \quad x = g(y) = (y - a)/b; \quad g'(y) = 1/b.$$

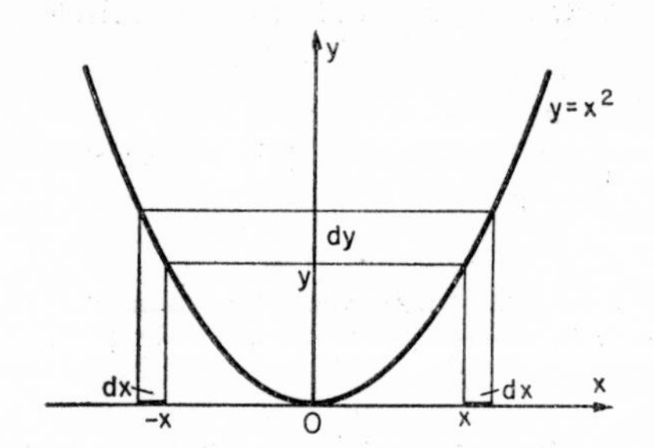

FIG. 13

The relationship existing between the probability density functions of ξ and η is therefore:

$$\psi(y) = \frac{1}{|b|}\varphi\left(\frac{y - a}{b}\right). \tag{2.32}$$

We leave the following to the reader: if the random variable ξ is uniformly distributed on the interval (α_1, α_2), show that the random variable $\eta = a + b\xi$ will be uniformly distributed on the interval $(a + b\alpha_1, a + b\alpha_2)$.

If ξ_0 is a standard normal variable with density function (2.22), then the random variable $\eta = a + b\xi_0$ will have the general normal distribution with density

$$\psi(y) = \frac{1}{|b|\sqrt{(2\pi)}} \exp\left(-\frac{(y - a)^2}{2b^2}\right).$$

This enables us to calculate probabilities for the general normal distribution (2.26) using the normal integral (2.23). In fact, let ξ be a random variable with the general normal distribution: then the random variable $\xi_0 = (\xi - a)/\sigma$ will have the standard normal distribution (2.22). Thus the inequality $x_1 < \xi < x_2$ is equivalent to the inequality

$$\frac{x_1 - a}{\sigma} < \xi_0 < \frac{x_2 - a}{\sigma},$$

so that we obtain

$$\mathbf{P}\{x_1 < \xi < x_2\} = \mathbf{P}\left\{\frac{x_1 - a}{\sigma} < \xi_0 < \frac{x_2 - a}{\sigma}\right\} = \tfrac{1}{2}\,\Phi(t_2) - \tfrac{1}{2}\,\Phi(t_1), \tag{2.33}$$

where

$$t_1 = \frac{x_1 - a}{\sigma}; \quad t_2 = \frac{x_2 - a}{\sigma}.$$

In the above we have treated only monotone functions. As an example of a non-monotone function we shall consider $\eta = \xi^2$ (where we will suppose that the range of ξ is the entire real line). Here $y = f(x) = x^2 \geqq 0$; the inverse function has two equivalent branches: $x = g_1(y) = +\sqrt{y}$, and $x = g_2(y) = -\sqrt{y}$ (see Fig. 13). Applying (2.31) to each of these branches and adding the terms with the same value of y, we derive, for $y > 0$:

$$\psi(y) = \varphi(\sqrt{y}) \left|\frac{1}{2\sqrt{y}}\right| + \varphi(-\sqrt{y}) \left|\frac{-1}{2\sqrt{y}}\right| = [\varphi(\sqrt{y}) + \varphi(-\sqrt{y})] \frac{1}{2\sqrt{y}}. \tag{2.34}$$

For $y < 0$ we must have $\psi(y) = 0$.

The idea of two-dimensional random variables. Functions of two random variables

For the study of functions of several random variables and also for the solution of many practical problems it is necessary to consider many-dimensional random variables, i.e. random variables whose values are distributed in a space of two, three, or more dimensions. As an example of a two-dimensional random variable we can take the point of impact of a shot on a target. If the coordinates of this point in the plane of the target are denoted by ξ and η, then the pair (ξ, η) is a two-dimensional random variable.

An example of a two-dimensional discrete random variable was discussed in Ex. 2, § 5 (p. 17); table (1.19) is an example of a probability distribution table for a two-dimensional discrete random variable. In this book it will not be possible to discuss many-dimensional random variables in detail. We will indicate only some expressions involving two-dimensional continuous random variables (the corresponding formulae for two-dimensional discrete random variables will have an analogous form).

A value of the random variable (ξ, η) is a point (x, y); the probability distribution is given by the probability differential

$$dP_{xy} = \varphi(x, y)\, dx\, dy, \tag{2.35}$$

which is the principal part of the probability that (ξ, η) takes a value in the rectangle $\begin{bmatrix} x < \xi < x + dx \\ y < \eta < y + dy \end{bmatrix}$ (see Fig. 14). The function $\varphi(x, y)$ is called *a two-dimensional probability density function.*

The probability that (ξ, η) takes a value in some region (D) is determined by the double integral

$$\mathbf{P}\{(\xi, \eta) \in (D)\} = \iint_{(D)} \varphi(x, y)\, dx\, dy.$$

The density $\varphi(x, y)$ can be any non-negative function satisfying the condition

$$\int\int \varphi(x, y)\, dx\, dy = 1,$$

where the integral is taken over the entire range of possible values of the random variable (ξ, η).

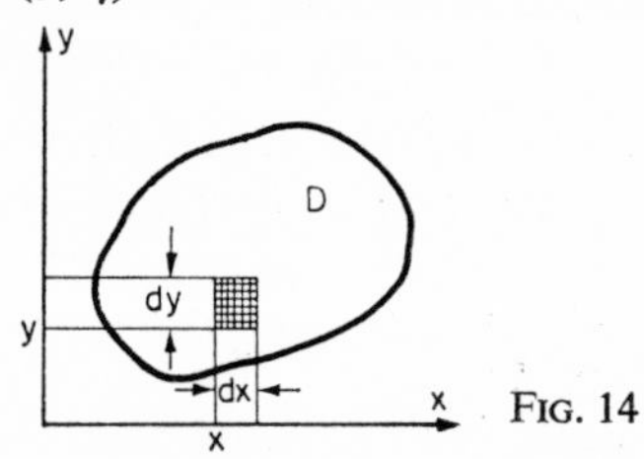

FIG. 14

The simplest example of a two-dimensional continuous random variable is *the random variable* (ξ, η) *with the uniform distribution on some finite region* (D_0). For this random variable the probability of some region (D) lying inside (D_0) is proportional to the area S_D of this region. Thus

$$dP_{xy} = \lambda\, dx\, dy \text{ for points inside } (D_0),$$

$$dP_{xy} = 0 \qquad \text{for points outside } (D_0).$$

We obtain the coefficient of proportionality, λ, from the condition

$$\mathbf{P}\{(\xi, \eta) \in (D_0)\} = \lambda S_{D_0} = 1,$$

so that $\lambda = 1/S_{D_0}$. Hence, the probability of (ξ, η) taking a value in the region (D) lying inside (D_0) is equal to the ratio of the areas S_D to S_{D_0}. The two-dimensional density is here equal to

$$\varphi(x, y) = \frac{1}{S_{D_0}}.$$

The coordinates ξ and η of a two-dimensional continuous random variable are themselves one-dimensional random variables.

The densities $\psi_1(x)$ and $\psi_2(x)$ of the random variables ξ and η are derived from the two-dimensional density $\varphi(x, y)$ by the following formulae:

$$\psi_1(x) = \int \varphi(x, y)\, dy, \tag{2.36}$$

$$\psi_2(y) = \int \varphi(x, y)\, dx. \tag{2.37}$$

The functions $\psi_1(x)$ and $\psi_2(y)$ are called *the marginal density functions.* To prove (2.36) it is sufficient to observe that the probability differential $\psi_1(x)\, dx$ is the probability of (ξ, η) taking a value in the hatched strip shown in Fig. 15, and therefore

$$\psi_1(x)\, dx = \left[\int \varphi(x, y)\, dy\right] dx.$$

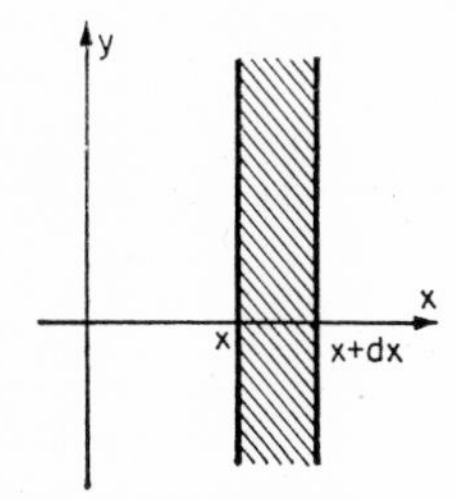

FIG. 15

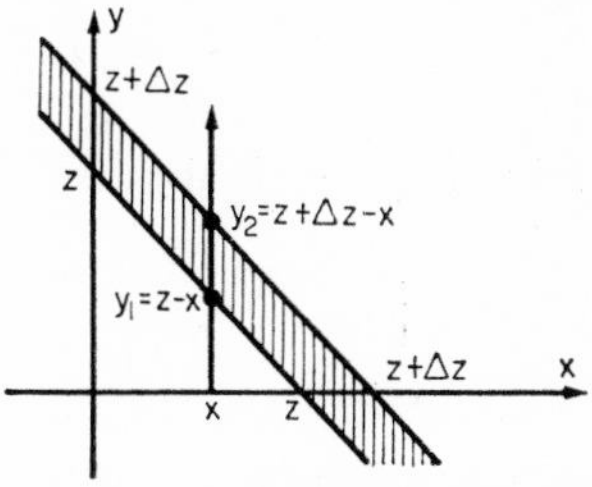

FIG. 16

For a discrete random variable with distribution (1.19) the analogous expression is (1.20).

The random variables ξ and η are called independent if the probability differential dP_{xy} is equal to the product of the probability differentials $\psi_1(x)\, dx$ and $\psi_2(y)\, dy$; i.e., if the two-dimensional density $\varphi(x, y)$ is equal to the product of the two one-dimensional densities:

$$\varphi(x, y) = \psi_1(x)\, \psi_2(y). \tag{2.38}$$

For a discrete random variable with distribution (1.19) the analogous expression is (2.8).

The probability distribution of the function $\zeta = f(\xi, \eta)$ of two random variables ξ and η is determined by the formula

$$\mathbf{P}\{z < \zeta < z + \Delta z\} = \iint_{(D_z, \Delta z)} \varphi(x, y)\, dx\, dy, \tag{2.39}$$

where $(D_{z,\Delta z})$ is the region of the xy-plane in which

$$z < f(x, y) < z + \Delta z,$$

and $\varphi(x, y)$ is the probability density function of the two-dimensional random variable (ξ, η). We can obtain the probability differential dP_z, and therefore the probability density function of $\zeta = f(\xi, \eta)$, from the integral (2.39) by evaluating the principal term of first order in Δz.

EXAMPLE. The distribution of the sum of random variables. For the sum $\zeta = \xi + \eta$ the region $(D_{z,\Delta z})$ is the strip included between the lines $x + y = z$ and $x + y = z + \Delta z$ (see Fig. 16). Therefore

$$\mathbf{P}\{z < \zeta < z + \Delta z\} = \int_{-\infty}^{\infty} dx \int_{z-x}^{z+\Delta z-x} \varphi(x, y)\, dy.$$

Evaluating the principal part of the inner integral we obtain

$$\int_{z-x}^{(z-x)+\Delta z} \varphi(x, y)\, dy \approx \varphi(x, z - x)\, \Delta z,$$

from which we derive the probability differential of the random variable ζ:

$$dP_z = \int_{-\infty}^{\infty} \varphi(x, z - x)\, dx\, \Delta z.$$

Thus we find the density function, $\chi(z)$, for the sum $\zeta = \xi + \eta$:

$$\chi(z) = \int_{-\infty}^{\infty} \varphi(x, z - x)\, dx. \qquad (2.40)$$

The sum of independent random variables is of particular interest. In this case we introduce (2.38) into (2.40) to obtain the probability density function of the sum $\zeta = \xi + \eta$ in the form

$$\chi(z) = \int_{-\infty}^{\infty} \psi_1(x)\, \psi_2(z - x)\, dx. \qquad (2.41)$$

The integral $\int_{-\infty}^{\infty} \psi_1(x)\, \psi_2(z - x)\, dx$ is called *the convolution* of the functions ψ_1 and ψ_2, and is denoted by $\psi_1 * \psi_2$.

EXERCISES

(1) Five balls are drawn from an urn containing 20 black and 4 white balls. Find the probability distribution of ξ the number of white balls drawn.

(2) From an urn containing 20 black and 4 white balls, balls are drawn consecutively until a black ball is drawn. Find the distribution of ξ, the number of white balls drawn before the first black balls.

(3) In a batch of n components there are k defectives. Determine the probability that in a sample of m chosen at random from the batch there will be l defective components.

(4) $\xi_1, \xi_2, \xi_3, \ldots$ is a sequence of independent random variables taking the values $+1$ or -1 with probabilities $\frac{1}{2}$ (compare ξ_k with the λ_k of (2.9)). If $\zeta_0 = 0$, $\zeta_n = \zeta_0 + \xi_1 + \cdots + \xi_n$, and $g_n = \mathbf{P}\{\zeta_n = 0\}$, show that for all $n \geqq 0$,

$$g_{2n+1} = 0, \qquad g_{2n} = \binom{2n}{n} 2^{-2n}. \tag{2.42}$$

(5) A trial consists in throwing three unbiased dice. Find the probability distribution for the sum of the three numbers which appear. Verify that it is more probable to get a sum of 11 than a sum of 12, although both 11 and 12 arise from six combinations:

$$11 = (6+4+1), (6+3+2), (5+5+1), (5+4+2), (5+3+3), (4+4+3);$$

$$12 = (6+5+1), (6+4+2), (6+3+3), (5+5+2), (5+4+3), (4+4+4).$$

(6) *The Multinomial Distribution.* We may generalize the binomial distribution in the following way. An experiment is performed in which each trial has k possible outcomes $A_1, A_2, \ldots, A_k$ with probabilities $p_1, p_2, \ldots, p_k$ respectively ($\Sigma p_i = 1$). Prove that, in a sequence of n independent trials the probability that A_1 occurs n_1 times, A_2 occurs n_2 times, ..., A_k occurs n_k times ($\Sigma n_i = n$), is

$$\frac{n!}{n_1!\, n_2! \ldots n_k!} p_1^{n_1} p_2^{n_2} \ldots p_k^{n_k}. \tag{2.43}$$

Construct an urn scheme for this experiment.

(7) The number of calls, X, passing through a switchboard during a day has a Poisson distribution, $\mathbf{P}\{X = n\} = e^{-\lambda}\lambda^n/n!$ If the calls are independent, and if the probability that a call is connected is p for all callers, what is the probability distribution of connected calls during a day through this switchboard?

(8) Prove that the most probable value of the frequency μ_n is the integer m_0 satisfying the inequality $np + p - 1 \leqq m_0 \leqq np + p$; in particular, if $(n+1)p$ is an integer, then $m_0 = (n+1)p$ or $(n+1)p - 1$.

(9) The cumulative distribution function of a random variable ξ has the form

$$F(x) = A + B \tan^{-1}(x/a) \quad (-\infty < x < \infty).$$

Determine the coefficients A and B. Find the probability density function. (This is called the *Cauchy distribution.*)

(10) The two-dimensional random variable (ξ, η) has the probability density function $f(x, y) = (2\pi\sigma^2)^{-1} \exp\{-(x^2 + y^2)/2\sigma^2\}$. If we make the transformation to polar coordinates, $\xi = r \cos \theta$, $\eta = r \sin \theta$, show that the random variables r and θ are independent, and write down their probability density functions.

(11) Use the result of the preceding example to show that the random variable ξ/η has the Cauchy distribution.

(12) The random variable η is uniformly distributed on the circumference of a circle of radius R. Taking a fixed diameter as the x-axis, find the probability distribution of the projection ξ of η onto this diameter.

(13) In a process for manufacturing metal cubes, the machine tool is set once for each cube and the 3 faces of the cube are then cut with this setting. If we assume that there is no change in the setting during the cutting of a cube, and if we assume that the setting is distributed with the general normal distribution (2.26), find the probability density function of the volume, v, of a cube.

(14) If ξ and η are independent and normally distributed with zero mean and unit variance, use the result of p.43 to evaluate the probability density function of $\xi^2 + \eta^2$.

(15) Find the distribution of the sum of two independent random variables, each uniformly distributed on the interval $(-1, +1)$.

(16) If ξ and η are independent random variables with Poisson distributions (parameters a and b respectively), show that $\xi + \eta$ is distributed according to the Poisson distribution with parameter $a + b$.

(17) If ξ and η are independent, standard normal variables, show directly, using (2.41), that $\xi + \eta$ has the density (2.26) with $a = 0$ and $\sigma = \sqrt{2}$. (For an indirect proof of this result see § 17.)

(18) Two people A and B agree to meet at a certain place between the hours of 0 and 1. The first one to arrive will wait 20 minutes for the other, and then he will go. Find the probability p that A and B will meet if we assume that their times of arrival are independent and each is uniformly distributed on the interval $(0, 1)$.

Probability generating functions

If ξ is a random variable taking integer values we may define a useful auxiliary function: the so-called probability generating function (p.g.f.). Writing $\mathbf{P}\{\xi = n\} = p_n$, we define the p.g.f., $\pi(u) = \Sigma p_n u^n$. Clearly $\pi(1) = 1$, and the series converges for $u < 1$.

(19) Obtain the p.g.f.s of (a) the binomial distribution (2.12); (b) the *geometric distribution* (2.4); (c) the Poisson distribution (2.5).

(20) If ξ_1 and ξ_2 are independent random variables with p.g.f.s $\pi_1(u) = \Sigma p_n^{(1)} u^n$ and $\pi_2(u) = \Sigma p_n^{(2)} u^n$ respectively, show that the p.g.f. of $\xi_1 + \xi_2$ is $\pi_1(u)\,\pi_2(u)$. Use this to re-derive the result of Exercise (16).

(21) If ξ is a random variable with p.g.f. $\pi(u)$, evaluate the p.g.f. of (a) 2ξ, (b) $-\xi$. If ξ takes non-negative values and $r_n = \mathbf{P}\{\xi \geqq n\}$, show that $\varrho(u) = \sum_{n=0}^{\infty} r_n u^n = \dfrac{1 - \pi(u)}{1 - u}$.

(22) *Continuation of Example* 3 (p. 21). Using Exercise (19b) show that the

hunter will use $n+r$ shots up to and including the rth hit with probability $\binom{n+r-1}{n} p^r q^n$. This is called the *negative binomial distribution.*

(23) Suppose that the number of taxis arriving at the taxi-rank of a large station in time t is a random variable ξ_t with Poisson distribution $\mathbf{P}\{\xi_t = n\} = e^{-at}(at)^n/n!$ If the number of passengers emerging from the station in time t and requiring taxis is a random variable η_t (independent of ξ_t) with the same Poisson distribution, derive the distribution of the excess of taxis over passengers at the rank at time t.

(24) *Continuation of* (4). Although the numbers g_n in (2.42) do not constitute a distribution, we may still define a generating function $\gamma(u) = \Sigma g_n u^n$. Show that $\gamma(u) = (1 - u^2)^{-1/2}$.

(25) *Continuation.* Denote by f_{2n} the probability that the sequence $\zeta_1, \zeta_2, \ldots, \zeta_k, \ldots$ returns to zero for the first time at $k = 2n$. Show that

$$g_{2n} = f_{2n} + f_{2n-2}g_2 + \cdots + f_2 g_{2n-2}.$$

We have $g_0 = 1$, and we take $f_0 = 0$. Writing $\varphi(u) = \Sigma f_{2n}u^{2n}$, obtain

$$\varphi(u) = 1 - (1 - u^2)^{1/2}, \tag{2.44}$$

and hence f_{2n}.

CHAPTER III

NUMERICAL CHARACTERISTICS OF PROBABILITY DISTRIBUTIONS

IN CALCULATIONS with discrete or continuous random variables it is not always worth-while to work with the probability tables or the density functions. Apart from the fact that the tables and density functions are not always precisely known, calculations with them are often complicated and cumbersome. It happens that we can solve many important practical problems using a few average values which are characteristic of the distribution. We study first the operation of Expectation, with which we construct these characteristics.

§ 10. MEAN VALUE. THE MATHEMATICAL EXPECTATION OF A RANDOM VARIABLE

We start with the simpler idea of an arithmetic mean. Suppose we have N objects (e.g., valves, the set of rainy days in a year) to each of which is attached a characteristic number (e.g., the time since installation of a valve, the amount of precipitation on a rainy day).

The arithmetic mean of these numbers is the sum of all the numbers divided by the total number of objects.

Denote by $x_1, x_2, \ldots, x_\nu$ the different values of the numerical characteristic. Let M_k be the number of elements having the value x_k $(k = 1, 2, \ldots)$, so that $N = M_1 + M_2 + \cdots + M_\nu$. The then arithmetic mean $\bar{x}$ is given by

$$\bar{x} = \frac{x_1 M_1 + x_2 M_2 + \cdots + x_\nu M_\nu}{N},$$

which we write also in the form

$$\bar{x} = x_1 \frac{M_1}{N} + x_2 \frac{M_2}{N} + \cdots + x_\nu \frac{M_\nu}{N}. \tag{3.1}$$

From this last expression it is clear that the arithmetic mean does not depend on the actual numbers $M_1, M_2, \ldots, M_\nu$, but only on the relative frequencies $(M_1/N), (M_2/N), \ldots, (M_\nu/N)$.

Turning now to random variables we look first at a particular discrete variable. Let us choose one object at random from the set of N objects described above.† Then the numerical characteristic of the object drawn is a discrete random variable with the following probability distribution table:

ξ	x_1	x_2	...	x_k	...	x_ν
	$\frac{M_1}{N}$	$\frac{M_2}{N}$	...	$\frac{M_k}{N}$	...	$\frac{M_\nu}{N}$

(3.2)

(since the conditions for the classical definition are satisfied, so that the probability of the value x_k is equal to the relative frequency of the objects carrying that value). The arithmetic mean of the characteristic number becomes here the "expectation" of the random variable ξ. This mean value is called the *mathematical expectation* of the random variable ξ and we shall denote it by $\mathbf{E}\xi$. Thus the mathematical expectation of the random variable (3.2) is equal to

$$\mathbf{E}\xi = x_1 \frac{M_1}{N} + x_2 \frac{M_2}{N} + \cdots + x_\nu \frac{M_\nu}{N},$$

where this sum must now be interpreted as the sum of the products of the values of ξ and their probabilities. This enables us at once to extend the idea of mathematical expectation to an arbitrary discrete random variable

ξ	x_1	x_2	...
	p_1	p_2	... .

† As in § 2 it is simplest to represent this trial as random sampling from an urn containing N balls, M_1 of which have a mark "x_1", M_2 have a mark "x_2", etc.

DEFINITION 1. *The mathematical expectation*, $\mathbf{E}\xi$, *of the discrete random variable* ξ *is defined to be the sum of products*

$$\mathbf{E}\xi = x_1 p_1 + x_2 p_2 + \cdots$$

or

$$\mathbf{E}\xi = \sum x_k p_k \tag{3.3}$$

where the sum is taken over the range of possible values of ξ. If the range is infinite then we will assume that the infinite series (3.3) is absolutely convergent (otherwise we shall say that $\mathbf{E}\xi$ does not exist; we shall not be concerned here with such random variables).

In seeking the definition of mathematical expectation for continuous random variables we recall that the probability differential $dP_x = \varphi(x)\, dx$ is the analogue of the probabilities p_k.

DEFINITION 2. *The mathematical expectation*, $\mathbf{E}\xi$, *of the continuous random variable* ξ *whose probability density function is* $\varphi(x)$ *is defined by*

$$\mathbf{E}\xi = \int x\, \varphi(x)\, dx, \tag{3.4}$$

where the integral is taken over the range of possible values of ξ. This integral is often written in the form $\int_{-\infty}^{\infty} x\varphi(x)\, dx$ even when the range of ξ is a finite interval: in this case we assume $\varphi(x) = 0$ outside the interval. If the range of ξ is an infinite interval then we assume that the improper integral $\int_{-\infty}^{\infty} x\varphi(x)\, dx$ is absolutely convergent (otherwise we say that $\mathbf{E}\xi$ does not exist; here also we shall not be concerned with such random variables).

We shall see below that all the properties of mathematical expectation are the same both for discrete and for continuous random variables.†

Properties of mathematical expectation

The most important property of the expectation is its linearity: *the mathematical expectation of a linear combination of random*

† It is possible to give a unified definition of the mathematical expectation of a random variable, but this requires some knowledge of the Stieltjes integral which is beyond the scope of this book.

variables is equal to the linear combination of their mathematical expectations:

$$\mathbf{E}(C_1\xi_1 + C_2\xi_2 + \cdots + C_n\xi_n) = C_1\mathbf{E}\xi_1 + C_2\mathbf{E}\xi_2 + \cdots + C_n\mathbf{E}\xi_n, \tag{3.5}$$

where $C_1, C_2, \ldots, C_n$ are constants.

For the proof of this relation it is sufficient to prove the following theorems.

(1) If C is a constant, then

$$\mathbf{E}C\xi = C\mathbf{E}\xi. \tag{3.6}$$

(2) The mathematical expectation of the sum of two random variables is the sum of their mathematical expectations (*the addition theorem for mathematical expectations*):

$$\mathbf{E}(\xi + \eta) = \mathbf{E}\xi + \mathbf{E}\eta. \tag{3.7}$$

Formula (3.6) is most simply proved for a discrete random variable. Multiplying by a constant C we obtain table (2.6) so that

$$\mathbf{E}C\xi = \sum (Cx_k)\, p_k = C \sum x_k p_k = C\mathbf{E}\xi.$$

We give the proof below (p. 54) for continuous random variables.

We prove the addition theorem (3.7) for continuous random variables ξ and η. If $\varphi(x, y)$ is the joint probability density function and $\chi(z)$ is the density function of the sum $\zeta = \xi + \eta$, then using (3.4) and (2.40) we have

$$\mathbf{E}\zeta = \int z\chi(z)\, dz = \int_{-\infty}^{\infty} z\, dz \int_{-\infty}^{\infty} \varphi(x, z - x)\, dx.$$

Changing the order of integration and replacing z by $x + y$ gives

$$\mathbf{E}(\xi + \eta) = \int_{-\infty}^{\infty} dx \int_{-\infty}^{\infty} z\varphi(x, z - x)\, dz$$

$$= \int_{-\infty}^{\infty} dx \int_{-\infty}^{\infty} (x + y)\, \varphi(x, y)\, dy. \tag{3.8}$$

If we now make use of the linearity of the integral and introduce expressions (2.36) and (2.37) we obtain

$$\mathbf{E}(\xi + \eta) = \int_{-\infty}^{\infty} x\,dx \int_{-\infty}^{\infty} \varphi(x, y)\,dy + \int_{-\infty}^{\infty} y\,dy \int_{-\infty}^{\infty} \varphi(x, y)\,dx$$

$$= \int_{-\infty}^{\infty} x\,\psi_1(x)\,dx + \int_{-\infty}^{\infty} y\,\psi_2(y)\,dy = \mathbf{E}\xi + \mathbf{E}\eta.$$

The proof goes through in the same way for discrete random variables. First of all, with the help of the auxiliary table (2.7), it is easy to see that the mathematical expectation of the sum $\xi + \eta$ can be expressed as

$$\mathbf{E}(\xi + \eta) = \sum_{k,l} (x_k + y_l)\,p_{kl},$$

where the sum is taken over the ranges of the random variables ξ and η. If we now write this sum in two parts as

$$\sum_{k,l} (x_k + y_l)\,p_{kl} = \sum_{k,l} x_k p_{kl} + \sum_{k,l} y_l p_{kl}$$

we can introduce (1.20) into the first part to derive

$$\sum_{k,l} x_k p_{kl} = \sum_{k} x_k \sum_{l} p_{kl} = \sum_{k} x_k p_k,$$

where $p_k = \mathbf{P}\{\xi = x_k\}$. Thus $\sum_{k,l} x_k p_{kl} = \mathbf{E}\xi$, and in the same way $\sum_{k,l} y_l p_{kl} = \mathbf{E}\eta$, so that (3.7) is proved.

We note two further properties of mathematical expectation.

(3) The mathematical expectation of a constant C is equal to this same constant.

In fact, we can consider the constant C as a random variable taking the one value C with probability 1. Then $\mathbf{E}C = C\,.\,1 = C$.

(4) The mathematical expectation of the product of two independent random variables is equal to the product of their mathematical expectations (*multiplication theorem for mathematical expectations*):

$$\mathbf{E}\xi\eta = \mathbf{E}\xi\mathbf{E}\eta \quad \text{for independent } \xi, \eta. \tag{3.9}$$

We carry out the proof for discrete random variables. Since ξ and η are independent the probability distribution of the product $(\xi\eta)$ is determined from the auxiliary table

x_1y_1	x_1y_2	x_2y_1	x_2y_2	...
p_1q_1	p_1q_2	p_2q_1	p_2q_2	...

by uniting equal values in the first line and adding the corresponding probabilities (cf. the procedure for the sum of random variables on p. 23). The mathematical expectation of the product $(\xi\eta)$ can therefore be written in the form

$$\mathbf{E}\xi\eta = \sum_{k,l} (x_k y_l)\, p_k q_l,$$

where the double sum is taken over the ranges of the random variables ξ and η. Rearranging this sum we obtain

$$\mathbf{E}\xi\eta = \sum_k \sum_l x_k y_l p_k q_l = \sum_k x_k p_k \sum_l y_l q_l,$$

and (3.9) is proved.

The proof of this expression for continuous random variables is easily carried out with the help of (3.13), and we leave it to the reader.

There is no difficulty in generalising (3.9) to an arbitrary number of independent factors.

Calculation of the mathematical expectation of a function

(1) Let ξ be a discrete random variable taking values x_k with probability p_k. The function $f(\xi)$ is again a discrete random variable and its mathematical expectation is given by the expression

$$\mathbf{E}f(\xi) = \sum f(x_k)\, \mathbf{P}\{f(\xi) = f(x_k)\}, \tag{3.10}$$

where the sum is taken over the different values of $f(x_k)$.

Now it turns out that we do not need to introduce the distribution of $f(\xi)$ in order to calculate its expectation: we can work directly with the distribution of the variable ξ. In fact, we have

$$\mathbf{E}f(\xi) = \sum f(x_k) p_k, \tag{3.11}$$

where the sum is taken over the range of the random variable ξ.

Before proving (3.11) we remark that if all the values $f(x_k)$ are different then the function $f(\xi)$ has the distribution table (2.28), and (3.11) coincides with (3.10). In the general case the values of $f(x_k)$ need not all be different. We shall suppose that $f(x_2) = f(x_1)$ and that all the other $f(x_k)$ are distinct. Then the probability of the event $(f(\xi) = f(x_1))$ is equal to $p_1 + p_2$, and the corresponding term in the sum (3.10) is therefore

$$f(x_1)\,\mathbf{P}\{f(\xi) = f(x_1)\} = f(x_1)\,(p_1 + p_2) = f(x_1)p_1 + f(x_2)p_2,$$

which gives us (3.11).

(2) When ξ is a continuous random variable the mathematical expectation of the function $f(\xi)$ can be obtained directly using the probability density function $\varphi(x)$ of the random variable ξ. We shall show that

$$\mathbf{E}f(\xi) = \int f(x)\,\varphi(x)\,dx. \tag{3.12}$$

We shall establish (3.12) only for the case when $f(x)$ is an increasing function. Let $\psi(y)$ be the probability density function of the random variable $\eta = f(\xi)$. Then from the definition of mathematical expectation

$$\mathbf{E}\eta = \int y\,\psi(y)\,dy.$$

If we make the substitution $y = f(x)$ it follows from (2.29) that $\psi(y)\,dy = \varphi(x)\,dx$ and we obtain (3.12) immediately. For example, if $f(\xi) = C\xi$, the expression (3.12) gives

$$\mathbf{E}C\xi = \int Cx\varphi(x)\,dx = C\int x\varphi(x)\,dx = C\mathbf{E}\xi,$$

which proves (3.6) for a continuous random variable.

(3) We quote without proof the corresponding results for calculating the mathematical expectation of a function of two variables.

When ξ and η are discrete,

$$\mathbf{E}f(\xi, \eta) = \sum_{k,l} f(x_k, y_l)p_{kl},$$

where the sum is taken over the ranges of the variables ξ and η, and p_{k_l} is the probability of the intersection of the events $(\xi = x_k)$ and $(\eta = y_l)$.

In the continuous case

$$\mathbf{E}f(\xi, \eta) = \iint f(x, y)\,\varphi(x, y)\,dx\,dy, \tag{3.13}$$

where $\varphi(x, y)$ is the probability density function of the random variable (ξ, η).

We remark that we have already discussed some special cases of these formulae: for example, $f(x, y) = x + y$ (see (3.8)), and $f(x, y) = xy$.

§ 11. THE CENTRE OF THE PROBABILITY DISTRIBUTION OF A RANDOM VARIABLE

The mathematical expectation of a random variable gives a convenient measure of the location of the distribution. The mathematical expectation is a real number lying in the range of the random variable. Thus, if the range of ξ is the interval (α_1, α_2), then

$$\mathbf{P}\{\alpha_1 < \xi < \alpha_2\} = \int_{\alpha_1}^{\alpha_2} \varphi(x)\, dx = 1,$$

and from the inequality

$$\int_{\alpha_1}^{\alpha_2} \alpha_1 \varphi(x)\, dx < \int_{\alpha_1}^{\alpha_2} x\varphi(x)\, dx < \int_{\alpha_1}^{\alpha_2} \alpha_2 \varphi(x)\, dx$$

it follows that

$$\alpha_1 < \mathbf{E}\xi < \alpha_2.\dagger$$

In the next chapter we will show that arithmetic means of measurements of the quantity ξ are grouped near to the mathematical expectation of ξ. In order to emphasise the role of mathematical expectation as a basic measure of location of a random variable (distinct from its role in the abstract operation of expectation) we give the following definition:

The centre of the probability distribution of a random variable ξ is its mathematical expectation $\mathbf{E}\xi$.‡

To clarify the idea of the centre of a distribution as a measure of location we give some examples.

(1) Let ξ be the number of used cartridges discussed in Ex. 2 of § 6 (p. 20). From the distribution table (2.3) we obtain the expected number of used cartridges:

$$\mathbf{E}\xi = 1 \times 0{\cdot}8 + 2 \times 0{\cdot}16 + 3 \times 0{\cdot}04 = 1{\cdot}24.$$

It turns out not to be an integer. To see how to make use of this result let us suppose that the hunter makes 100 attempts under the same conditions (i.e. at each attempt he is allowed 3 cartridges

† In particular if ξ takes only positive values then $\mathbf{E}\xi > 0$.

‡ The analogous idea from mechanics of the centre of gravity of a distribution of mass will clarify the term "centre of a probability distribution": if, for example, masses $p_1, p_2, \ldots, p_\nu$ are concentrated in points $x_1, x_2, \ldots, x_\nu$ then the centre of gravity x_c is given by

$$x_c = \frac{x_1 p_1 + x_2 p_2 + \cdots + x_\nu p_\nu}{p_1 + p_2 + \cdots + p_\nu}.$$

If $p_1 + p_2 + \cdots + p_\nu = 1$, then this expression coincides with (2.3).

and table (2.3) holds). Let ξ_k be the number of cartridges used at the kth attempt; then

$$\xi = \xi_1 + \xi_2 + \cdots + \xi_{100}$$

is the total number of cartridges used in a hundred attempts. To calculate the expectation we use the linearity property and the fact that $\mathbf{E}\xi_k = 1{\cdot}24$ $(k = 1, 2, \ldots, 100)$:

$$\mathbf{E}\xi = \mathbf{E}\xi_1 + \mathbf{E}\xi_2 + \cdots + \mathbf{E}\xi_{100} = 100 \times 1{\cdot}24 = 124.$$

In practice this means that in 100 similar attempts the hunter will use, on the average, 124 cartridges.

(2) *The centre of the Poisson distribution* (2.5):

$$\mathbf{E}\xi = \sum_{m=0}^{\infty} m \frac{a^m}{m!} e^{-a}$$

$$= e^{-a} a \left(1 + a + \frac{a^2}{2!} + \cdots + \frac{a^{m-1}}{(m-1)!} + \cdots \right) = a.$$

The constant a occurring in the Poisson distribution (2.5) is therefore the mathematical expectation of a random variable ξ which has this distribution.

(3) *The centre of the Binomial distribution.* From the table (2.14) we can write down immediately

$$\mathbf{E}\mu_n = \sum_{m=0}^{n} m \binom{n}{m} p^m q^{n-m}.$$

However, the calculation is more quickly carried out using the representations (2.10) and (2.11) of the random variables μ_n and ω_n in terms of indicator random variables $\lambda_1, \lambda_2, \ldots, \lambda_n$. First of all, table (2.9) gives immediately:

$$\mathbf{E}\lambda_k = 1p + 0q = p \quad (k = 1, 2, \ldots, n). \tag{3.14}$$

In other words, the centre of the distribution of an indicator random variable is equal to the probability of the event whose occurrence or non-occurrence it indicates. Then we obtain

$$\mathbf{E}\mu_n = \mathbf{E}\lambda_1 + \mathbf{E}\lambda_2 + \cdots + \mathbf{E}\lambda_n = np, \tag{3.15}$$

and

$$\mathbf{E}\omega_n = \frac{1}{n} \mathbf{E}\mu_n = p. \tag{3.16}$$

Thus the centre of the distribution of the relative frequency ω_n of an event is equal to the probability of occurrence of this event in a single trial. The centre of the distribution of the frequency μ_n is equal to this probability n times. Now this completely agrees with our intuitive interpretation of mathematical expectation. For example, if the probability is equal to $p = 0{\cdot}2$ and the experiment is performed $n = 100$ times, then we expect that the event will occur $np = 20$ times. In precisely the same way if we are told that the probability of a defective in a large batch of manufactured articles is equal to $p = 1\% = 0{\cdot}01$, then if we take a sample of $n = 1000$ articles we would expect to discover $np = 10$ defectives. (We remark that we should expect some fluctuation but the question here concerns the number of times on the average we expect an event to occur). We note also that the derivation of (3.16) from (3.14) depends on the linearity of the mathematical expectation. If the event A has probability p_k at the kth trial, then the centre of the distribution of ω_n will be equal to

$$\mathbf{E}\omega_n = \frac{1}{n}(\mathbf{E}\lambda_1 + \mathbf{E}\lambda_2 + \cdots + \mathbf{E}\lambda_n) = \frac{p_1 + p_2 + \cdots + p_n}{n},$$

i.e. it is equal to the arithmetic mean of the probabilities of the event A in the n trials.

(4) *The centre of the uniform distribution.* If the random variable ξ is uniformly distributed on the interval (α_1, α_2) then the centre of the distribution coincides with the mid-point of the interval. In fact, the probability density is equal to $1/(\alpha_2 - \alpha_1)$ on the interval (α_1, α_2) so that

$$\mathbf{E}\xi = \int_{\alpha_1}^{\alpha_2} x \frac{1}{\alpha_2 - \alpha_1} dx = \frac{1}{\alpha_2 - \alpha_1} \cdot \frac{\alpha_2^2 - \alpha_1^2}{2} = \frac{\alpha_1 + \alpha_2}{2}.$$

(5) *The centre of the normal distribution.* The centre of the standard normal distribution (2.22) is equal to zero, because the density $\varphi_0(x)$ is an even function:

$$\mathbf{E}\xi_0 = \frac{1}{\sqrt{(2\pi)}} \int_{-\infty}^{\infty} x e^{-x^2/2} dx = 0.$$

A random variable ξ with the general normal distribution (2.26) can be expressed in terms of a standard normal variable ξ_0 (see p. 42):

$$\xi = a + \sigma\xi_0.$$

Consequently

$$\mathbf{E}\xi = a + \sigma \mathbf{E}\xi_0 = a.$$

Thus, the centre of the general normal distribution is equal to the parameter a (this establishes the meaning of this parameter and agrees with the symmetry shown by the normal curve about the line $x = a$: see Fig. 8).

Note: If the probability curve is symmetric about some line $x = a$ then the centre of the distribution coincides with the point a.

§ 12. THE MEASURE OF THE DISPERSION OF A RANDOM VARIABLE. THE MOMENTS OF A DISTRIBUTION

The dispersion of a random variable ξ will be measured by some function of the deviation $\xi - a$, where $a = \mathbf{E}\xi$, the centre of the distribution. If we take the expectation of this deviation as it stands, we do not find a numerical measure of the dispersion because

$$\mathbf{E}(\xi - a) = \mathbf{E}\xi - a = 0,$$

i.e. deviations with opposite signs cancel each other out.

The fundamental measure of the dispersion of a random variable ξ is the standard deviation† *σ defined by*

$$\sigma = \sigma(\xi) = \sqrt{[\mathbf{E}(\xi - a)^2]} \quad \text{where} \quad a = \mathbf{E}\xi. \tag{3.17}$$

The number $\mathbf{E}(\xi - a)^2$ is called the *variance*† of the random variable ξ and is denoted by $\mathbf{D}\xi$:

$$\mathbf{D}\xi = \mathbf{E}(\xi - a)^2 = \sigma^2(\xi).$$

Using (3.11) and (3.12) we can write the expression for the variance in the form

$$\sigma^2(\xi) = \sum (x_k - a)^2 p_k,$$

$$\sigma^2(\xi) = \int (x - a)^2 \varphi(x)\, dx,$$

† *Translator's note:* These are the common Anglo-American usages. The author uses the expressions "root mean square deviation" and "dispersion" respectively.

according as ξ is discrete or continuous.† Writing it in this way we see that the standard deviation has the same dimension taken by the random variable. The special role of the standard deviation as an estimate of the dispersion of the random variable will be discussed in detail below (principally in Chapters IV and V). In particular we will show that values of the random variable will not occur in practice which deviate from the centre of the distribution by more than some small multiple of σ. We limit our attention in this section to examples and the simplest properties of the standard deviation.

Fundamental properties of the standard deviation

(1) If ξ is a random variable and C is a constant, then

$$\sigma(C\xi) = |C|\,\sigma(\xi) \tag{3.18}$$

$$\sigma(C + \xi) = \sigma(\xi). \tag{3.19}$$

We prove these immediately as follows:

$$\begin{aligned}\sigma^2(C\xi) &= \mathbf{E}(C\xi - \mathbf{E}C\xi)^2 = \mathbf{E}(C\xi - Ca)^2 \\ &= C^2\mathbf{E}(\xi - a)^2 = C^2\sigma^2(\xi);\end{aligned}$$

$$\begin{aligned}\sigma^2(\xi + C) &= \mathbf{E}[(\xi + C) - \mathbf{E}(\xi + C)]^2 = \mathbf{E}[(\xi + C) - (a + C)]^2 \\ &= \mathbf{E}(\xi - a)^2 = \sigma^2(\xi).\end{aligned}$$

(2) If ξ and η are *independent* random variables then the variance of the sum is equal to the sum of the variances,

$$\sigma^2(\xi + \eta) = \sigma^2(\xi) + \sigma^2(\eta) \tag{3.20}$$

(*addition theorem for variances*), and therefore

$$\sigma(\xi + \eta) = \sqrt{[\sigma^2(\xi) + \sigma^2(\eta)]}.$$

† It is worth continuing the mechanical analogy begun in the footnote to p. 57. If we interpret $p_1, p_2, \ldots, p_\nu$ as masses concentrated on the x-axis in the points $x_1, x_2, \ldots, x_\nu$, then the variance $\sigma^2 = \sum (x_k - a)^2 p_k$ becomes the moment of inertia of this system of mass-points (with respect to the centre of gravity a).

We prove (3.20). We denote $\mathbf{E}\xi = a$, $\mathbf{E}\eta = b$, so that $\mathbf{E}(\xi + \eta) = a + b$. Then

$$\sigma^2(\xi + \eta) = \mathbf{E}[(\xi + \eta) - (a + b)]^2$$
$$= \mathbf{E}[(\xi - a)^2 + 2(\xi - a)(\eta - b) + (\eta - b)^2],$$

and invoking the linearity of the mathematical expectation we obtain $\sigma^2(\xi + \eta) = \mathbf{E}(\xi - a)^2 + 2\mathbf{E}(\xi - a)(\eta - b) + \mathbf{E}(\eta - b)^2$. Since the random variables ξ and η are independent we can apply the multiplication theorem for mathematical expectations:

$$\mathbf{E}(\xi - a)(\eta - b) = \mathbf{E}(\xi - a)\,\mathbf{E}(\eta - b).$$

Now we proved above that $\mathbf{E}(\xi - a) = 0$, so that

$$\mathbf{E}(\xi - a)(\eta - b) = 0$$

and (3.20) is proved.

The addition theorem for variances can be extended without difficulty to an arbitrary number of pair-wise independent random variables.

COROLLARY. The variance of a linear combination of pair-wise independent random variables $\xi_1, \xi_2, \ldots, \xi_n$ is given by the expression

$$\sigma^2(C_1\xi_1 + C_2\xi_2 + \cdots + C_n\xi_n)$$
$$= C_1^2\sigma^2(\xi_1) + C_2^2\sigma^2(\xi_2) + \cdots + C_n^2\sigma^2(\xi_n). \quad (3.20')$$

This follows immediately from (3.20) and (3.18).

As a particular case of this last relation suppose that the random variables $\xi_1, \xi_2, \ldots, \xi_n$ have the same variance

$$\sigma^2(\xi_k) = \sigma^2 \quad (k = 1, 2, \ldots, n),$$

then the variance of the arithmetic mean is equal to

$$\sigma^2\left(\frac{\xi_1 + \xi_2 + \cdots + \xi_n}{n}\right) = \frac{1}{n^2}[\sigma^2(\xi_1) + \sigma^2(\xi_2) + \cdots + \sigma^2(\xi_n)] = \frac{\sigma^2}{n}.$$

Consequently the standard deviation is given by

$$\sigma\left(\frac{\xi_1 + \xi_2 + \cdots + \xi_n}{n}\right) = \frac{\sigma}{\sqrt{n}}. \quad (3.21)$$

This last result plays an important part in the theory of observations.

EXAMPLE 1. *The standard deviation of the relative frequency.* According to (2.11) the relative frequency ω_n is the arithmetic mean of mutually independent indicator random variables $\lambda_1, \lambda_2, \ldots, \lambda_n$ identically distributed with distribution table (2.9):

$$\omega_n = \frac{\lambda_1 + \lambda_2 + \cdots + \lambda_n}{n}, \quad \text{where} \quad \lambda_k \begin{array}{c|c} 1 & 0 \\ \hline p & q \end{array}, \quad p + q = 1$$

$$(k = 1, 2, \ldots, n).$$

We can obtain the variance of λ_k immediately: if we recall that the centre of the distribution is p we have

$$\sigma^2(\lambda_k) = \mathbf{E}(\lambda_k - p)^2 = (1-p)^2 p + (0-p)^2 q = q^2 p + p^2 q = pq.$$

Hence we have the standard deviation:

$$\sigma(\lambda_k) = \sqrt{(pq)} \quad (k = 1, 2, \ldots, n).$$

Inserting this in (3.21) we derive

$$\sigma(\omega_n) = \sqrt{\frac{pq}{n}}. \tag{3.22}$$

Applying (3.18) to this we obtain also

$$\sigma(\mu_n) = \sigma(n\omega_n) = n\sigma(\omega_n) = \sqrt{(npq)}.$$

EXAMPLE 2. *The standard deviation of a random variable uniformly distributed on the interval* (α_1, α_2). We found the centre of this distribution:

$$a = \mathbf{E}\xi = \frac{\alpha_1 + \alpha_2}{2}.$$

We calculate the variance directly:

$$\sigma^2(\xi) = \mathbf{E}\left(\xi - \frac{\alpha_1 + \alpha_2}{2}\right)^2 = \int_{\alpha_1}^{\alpha_2} \left(x - \frac{\alpha_1 + \alpha_2}{2}\right)^2 \frac{1}{\alpha_2 - \alpha_1}\, dx$$

$$= \frac{(\alpha_2 - \alpha_1)^2}{12}.$$

Whence the standard deviation is

$$\sigma(\xi) = \frac{\alpha_2 - \alpha_1}{2\sqrt{3}}$$

proportional to the length of the interval (α_1, α_2) (and approximately one third of this length).

EXAMPLE 3. *The variance of the normal distribution.* If ξ_0 is a standard normal variable with distribution (2.22) the centre is $\mathbf{E}\xi_0 = 0$ and the variance is equal to

$$\mathbf{D}\xi_0 = \mathbf{E}\xi_0^2 = \frac{1}{\sqrt{(2\pi)}} \int_{-\infty}^{\infty} x^2 e^{-x^2/2} \, dx = 1.\dagger \quad (3.23)$$

The random variable $\xi = a + \sigma\xi_0$ has the general normal distribution (2.26), and its variance is equal to

$$\mathbf{D}\xi = \mathbf{D}(a + \sigma\xi_0) = \sigma^2 \mathbf{D}\xi_0 = \sigma^2.$$

Thus $\sigma(\xi) = \sigma$. We have already established that $\mathbf{E}\xi = a$. This last expresssion clarifies for us the meaning of the other parameter, σ, in the distribution (2.26). The centre of the general normal distribution is a and the variance is σ^2. Figure 7 (p. 37) clearly illustrates the role of σ as a measure of the dispersion of a random variable with the general normal distribution (for $a = 0$).

The minimal property of the centre

The variance of a random variable ξ possesses the following minimal property:

$$\mathbf{E}(\xi - a)^2 < \mathbf{E}(\xi - C)^2,$$

where $a = \mathbf{E}\xi$ and $C \neq a$.

† Integrating by parts we have

$$\frac{1}{\sqrt{(2\pi)}} \int_{-\infty}^{\infty} x \, xe^{-x^2/2} \, dx = \frac{1}{\sqrt{(2\pi)}} \left[-xe^{-x^2/2} \Big|_{-\infty}^{\infty} + \int_{-\infty}^{\infty} e^{-x^2/2} \, dx \right]$$

$$= \frac{1}{\sqrt{(2\pi)}} \int_{-\infty}^{\infty} e^{-x^2/2} \, dx = 1.$$

Proof: Because $\mathbf{E}(\xi - a) = 0$, we have

$$\begin{aligned}\mathbf{E}(\xi - C)^2 &= \mathbf{E}[(\xi - a) + (a - C)]^2 \\ &= \mathbf{E}(\xi - a)^2 + 2(a - C)\,\mathbf{E}(\xi - a) + (a - C)^2 \\ &= \mathbf{E}(\xi - a)^2 + (a - C)^2, \qquad (3.24)\end{aligned}$$

so that

$$\mathbf{E}(\xi - C)^2 \geqq \mathbf{E}(\xi - a)^2,$$

and this is an equality only when $(a - C)^2 = 0$, i.e. when $a = C$.

The property which we have just proved establishes an important relationship between the centre of a distribution and the variance: the centre of the distribution minimises the quantity $\mathbf{E}(\xi - C)^2$, and the minimum of this expected value is just the variance $\sigma^2(\xi)$.

It is often possible to make use of (3.24) in the calculation of variances.† In particular, if $C = 0$ it becomes

$$\sigma^2(\xi) = \mathbf{E}(\xi - a)^2 = \mathbf{E}\xi^2 - a^2. \qquad (3.25)$$

As an example we calculate the variance of the Poisson distribution (2.5). It is not difficult to calculate $\mathbf{E}\xi^2$:

$$\begin{aligned}\mathbf{E}\xi^2 &= \mathbf{E}\xi(\xi - 1) + \mathbf{E}\xi = \sum_{m=0}^{\infty} m(m - 1)\frac{a^m}{m!}e^{-a} + a \\ &= a^2 e^{-a} \sum_{m=2}^{\infty} \frac{a^{m-2}}{(m - 2)!} + a = a^2 + a,\end{aligned}$$

from which we find the variance

$$\sigma^2(\xi) = \mathbf{E}\xi^2 - a^2 = (a^2 + a) - a^2 = a.$$

Thus in the Poisson distribution both the centre of the distribution and the variance are equal to the parameter a.

The moments of a distribution

We have considered above the two fundamental characteristics of a distribution: the centre $\mathbf{E}\xi = a$, and the variance $\mathbf{E}(\xi - a)^2 = \sigma^2$. These are important cases of the moments of the distribu-

† One should compare (3.24) with the corresponding theorem for moments of inertia.

tion introduced by the well-known Russian mathematician P.L. Chebyshev.

The *primitive moment* of order k is the mathematical expectation of the kth power of the random variable, i.e. $\mathbf{E}\xi^k$.

The *central moment* of order k is the mathematical expectation of the kth power of the deviation from the centre, i.e. $\mathbf{E}(\xi - a)^k$.

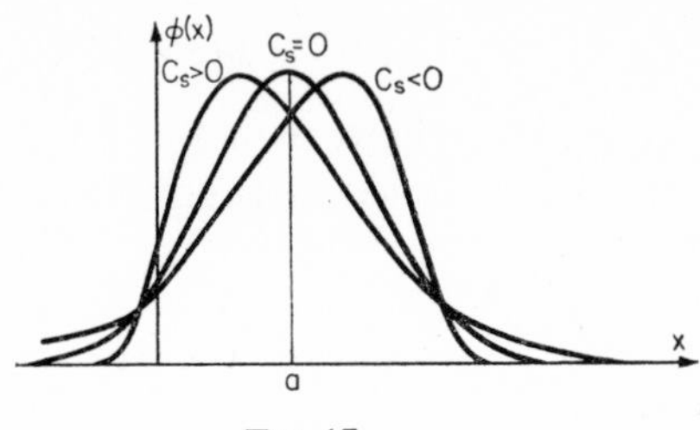

FIG. 17

There is a simple relation between the primitive and central moments which can be easily established using the binomial expansion. For example:

$$\mathbf{E}(\xi - a)^2 = \mathbf{E}\xi^2 - 2a\mathbf{E}\xi + a^2 = \mathbf{E}\xi^2 - a^2,$$

$$\mathbf{E}(\xi - a)^3 = \mathbf{E}\xi^3 - 3a\mathbf{E}\xi^2 + 3a^2\mathbf{E}\xi - a^3$$

$$= \mathbf{E}\xi^3 - 3a\mathbf{E}\xi^2 + 2a^3,$$

etc. The first of these expressions is (3.25).

We remarked above that $\mathbf{E}\xi$ and $\mathbf{E}(\xi - a)^2$ characterise the location and dispersion of the random variable ξ. The third order central moment $\mathbf{E}(\xi - a)^3$ is used to measure the asymmetry of the distribution. If the probability curve is symmetric about the line $x = a$ then the third order central moment (and in fact all central moments of odd order) will be zero.† Therefore, if $\mathbf{E}(\xi - a)^3$ differs from zero we can conclude that the distribution is not symmetric. It is usual to define the dimensionless *coefficient of skewness* $C_s = \dfrac{\mathbf{E}(\xi - a)^3}{\sigma^3(\xi)}$ as a measure of the asymmetry of the distribution. The sign of C_s indicates whether the probability curve is asymmetric to the right or left (see Fig. 17).

† This follows from the fact that if $\psi(y)$, the density function of $\eta = \xi - a$, is an even function then all products $y^{2k+1}\psi(y)$ will be odd functions.

Moments of higher orders do not occur in elementary problems of probability theory or in the simplest of its applications.

EXERCISES

(1) Evaluate the mathematical expectation of the product of the indicator random variables introduced on p. 30 for the scheme of sampling without replacement. Show that in this case the multiplication theorem for expectations does not apply.

(2) Using (3.13) prove the multiplication theorem for expectations when the random variables are continuous.

(3) Find the centre and standard deviation of the distribution (2.2).

(4) Find the centre and standard deviation of the sum of the numbers appearing when two dice are thrown.

(5) Find the mathematical expectation of the number of white balls in the scheme described in Exercise 1 of Chapter II.

(6) Find the centre and variance of the distribution (2.4) (Example 3, p. 21). Consider the case when $p = \frac{1}{10}$ and give an interpretation of the mathematical expectation.

(7) Find the centre and the standard deviation of the random variable $\xi_t - \eta_t$ discussed in Chapter II, Exercise 23. Comment on the values you obtain in terms of the physical situation it purports to describe.

(8) Find the mathematical expectation of the random variable $\zeta = \varphi(\xi, \eta) = +\sqrt{(\xi^2 + \eta^2)}$, where the two-dimensional random variable (ξ, η) has probability density function

$$f(x, y) = \begin{cases} \dfrac{1}{\pi R^2} & \text{for} \quad x^2 + y^2 \leqq R^2, \\ 0 & \text{for} \quad x^2 + y^2 > R^2. \end{cases}$$

(9) Find the centre and standard deviation of the random variable ξ discussed in Chapter II, Exercise 12.

(10) Find the centre and variance of the Pearson distribution (2.27).

(11) Show that, if the distribution function $F(x) = \mathbf{P}\{\xi < x\}$ satisfies the conditions

$$\lim_{x \to -\infty} [xF(x)] = 0, \qquad \lim_{x \to +\infty} \{x[1 - F(x)]\} = 0,$$

then the mathematical expectation of the random variable ξ is given by

$$\mathbf{E}\xi = \int_0^\infty [1 - F(x)]\, dx - \int_{-\infty}^0 F(x)\, dx.$$

(12) Find the mathematical expectation of the random variable v of Chapter II, Exercise 13.

(13) ξ and η are independent random variables. Prove the addition theorem for the central moment of third order

$$\mathbf{E}[(\xi + \eta) - (a + b)]^3 = \mathbf{E}(\xi - a)^3 + \mathbf{E}(\eta - b)^3.$$

(14) Find the coefficient of skewness of the binomial distribution.

(15) Prove that the coefficient of skewness of the Pearson distribution (2.27) is twice the so-called coefficient of variation

$$C_v = \frac{\sigma(\xi)}{\mathbf{E}\xi} = \frac{1}{\sqrt{\alpha}}.$$

(16) Calculate the fourth order central moment of the general normal distribution (2.26).

(17) $S = \frac{1}{2}ab \sin \gamma$ is the area of a triangle of sides a, b and included angle γ. If a, b are known without error and γ is uniformly distributed on the interval $(\gamma_0 - \alpha, \gamma_0 + \alpha)$, expand $\sin \gamma$ in a Taylor series about γ_0 to show that $\mathbf{E}\{S\} \approx \frac{1}{2}ab(1 + \frac{1}{6}\alpha^2) \sin \gamma_0$; compare with the exact value of $\mathbf{E}\{S\}$.

(18) Let ξ be a random variable with mathematical expectation μ, variance σ^2, and finite third moment. By developing the function $g(x)$ in a Taylor series about the point μ, show that $\mathbf{E}[g(\xi)] \approx g(\mu)$ and $\sigma^2[g(\xi)] \approx [g'(\mu)]^2\sigma^2$ provided $g''(\mu)$ is small and $g'''(x)$ is small over the range of values of ξ.

(19) If the random variable ξ has p.g.f. $\pi(u)$, show that $\mathbf{E}\xi = \pi'(1)$, $\mathbf{E}[\xi(\xi-1)] = \pi''(1)$, etc. (The quantity $\mathbf{E}[\xi(\xi-1)\ldots(\xi-n+1)]$ is called the nth *factorial moment.*) Use this and Chapter II, Exercise (19), to re-derive the centres and variances of (a) the binomial distribution; (b) the geometric distribution (cf. Exercise (6)); (c) the Poisson distribution.

(20) Show that the centre of the distribution generated by (2.44) is infinity.

CHAPTER IV

THE LAW OF LARGE NUMBERS

IT IS impossible to predict the value which a random variable will take as the result of a single trial. However, the sum of large number of random variables almost loses its random character and becomes constant. Because of the interaction of many sources of randomness, their joint effect becomes almost independent of chance. This phenomenon is given a precise meaning in theorems which are usually called Laws of Large Numbers.

§ 13. ON EVENTS WITH VERY SMALL PROBABILITY

We recall that the probability p of an event is a number measuring the possibility of its occurrence as a result of a certain experiment. The relative frequency of this event is the random variable ω_n, the centre of whose distribution is just the probability p (see § 11, example 3). When the value of the probability cannot be found directly from the experiment, each sample of n trials gives an experimental value of the relative frequency. We have remarked earlier that for a sufficiently large number, n, of trials the relative frequency ω_n turns out to be very near to the probability p. This general principle connects theory with practice, but it makes no statement about the number of trials which would be sufficient (we know that the relative frequency is near the probability but we do not know how near). In order to discuss this further we need a more definite criterion. We proceed to establish one using events with very small probabilities.

If an event has a very small probability then it will occur very rarely. For example, if an event has probability 0·000001 then it will occur approximately once in a million trials.†

† This does not mean that it will occur at the millionth trial; it can occur among the first trials.

Experience convinces us that an event with so small a probability will not as a rule occur at all in a small number of trials; we can therefore neglect the possibility of its occurring. For example, in a lottery of 1,000,000 tickets where there is one prize, it is doubtful if a particular ticket-holder can seriously reckon on the prize (although one of the million ticket-holders will certainly win the prize!). Can he any more reasonably reckon on it if there are 500,000 tickets, or 10,000? The question arises, how small must the probability of an event be in order that we may neglect the possibility of its occurrence in a single trial. The answer to this question cannot be given by the theory of probability but depends on the particular practical problem under consideration. We clarify this by means of an example in which we compare two events.

(a) In the automatic machining of components let the probability of producing a non-standard item be 0·01. Moreover a non-standard item is to be rejected at the assembly stage. If the part is inexpensive then we need not maintain a continuous control over all items before assembly; i.e. we neglect the probability 0·01 of an item being non-standard.

(b) In making parachutes the probability of producing one which will not open is 0·01. It is clear that in this case we cannot neglect a probability of 0·01 because this would result in the death of approximately every hundredth parachutist. In this case we must inspect every parachute produced.

In each application of probability theory we must prescribe what we mean by "very small" probabilities. We set a bound with respect to which we establish our criterion of practical impossibility: *we suppose that an event having probability less than the prescribed bound will not occur in a single trial.*

This criterion plays a fundamental role in the practical applications of probability theory. It enables us to see the practical significance of the theorems which we shall prove in the next sections.

This criterion is sometimes stated in a different way. The prescribed bound α is called a "confidence level": if the probability of an event A is less than α, then we are confident that in practice the event A will not occur in a single trial. It follows that the probability of the complementary event $\bar{A}$ is greater than $1-\alpha$, i.e. it will be near to one, just as the probability of A is near to zero. The confidence criterion can therefore be put in the alternative form: if the probability of an event is greater than $1-\alpha$, then we are practically confident that the event will occur (at a single trial).

We make one more remark about the above criterion. Let us suppose that according to some hypothesis the probability of an event A is very small (less than the prescribed bound). However, when we perform the experiment, the event A occurs. Then it is reasonable to doubt our initial hypothesis and to seek some other cause for the occurrence of the event A. This point is nicely illustrated by an old anectode quoted by Bertrand in his book *The Calculus of Probabilities* (1889). "One day in Naples the abbé Galliani met a man from Basilicata who, shaking three dice in a cup, bet that he would throw three sixes: and in fact he immediately obtained three sixes. You tell me how a man can be so lucky.† Now the man from Basilicata repeated his bet, and won a second time.‡ He put the dice back in the cup three, four, and five times, and each time he threw three sixes. 'The devil take it' cried the abbé, 'Those dice must be filled with lead'. And that was how it was."

§ 14. BERNOULLI'S THEOREM AND THE STABILITY OF RELATIVE FREQUENCIES

In some trial let us suppose that an event A has probability p. Let us suppose also that the trial is repeated n times.†† Then we know that the relative frequency of the event is a random variable ω_n whose mathematical expectation is equal to p. As for the standard deviation of ω_n, (3.22) shows that it decreases as the number n of trials increases:

$$\sigma(\omega_n) = \frac{\sqrt{(pq)}}{\sqrt{n}}.$$

Thus as the number of trials increases the values of the relative frequency of the event will become less spread out, i.e. they will become more closely grouped near the probability of the event.

Bernoulli's remarkable theorem was published in 1713.

BERNOULLI'S THEOREM. *If throughout a sequence of independent trials the probability, p, of an event remains unchanged, then the probability that the difference between ω_n and p exceeds a given number $\varepsilon > 0$ tends to zero as the number of trials increases without limit:*

$$\lim_{n\to\infty} \mathbf{P}\{|\omega_n - p| > \varepsilon\} = 0. \tag{4.1}$$

† Assuming the dice are unbiased the probability of this is $1/6^3 = 1/216$.
‡ Under the same conditions the probability of this is $1/(216)^2 = 1/46{,}656$.
†† See footnote, p. 25.

The fact that the probability of the inequality

$$|\omega_n - p| > \varepsilon$$

tends to zero means that for sufficiently large n this probability will become less than some prescribed upper bound (our criterion for very small probability: see § 13). Thus we will have confidence that in practice the above inequality will not be satisfied, and consequently that the complementary inequality

$$|\omega_n - p| \leqq \varepsilon \tag{4.2}$$

will be satisfied.

Bernoulli's theorem can be formulated alternatively as: *For a sufficiently large number of trials we may be confident that the difference between the relative frequency of an event and its probability will not exceed in absolute value some previously specified number* ε *(which may be arbitrarily small).*

Bernoulli's theorem is a very special case of the theorem of Chebyshev which we will prove in the next section. We remark that however large n may be we cannot categorically assert that inequality (4.2) will be satisfied: we can only have confidence that in practice (in the sense of § 13) this inequality will be satisfied. In order to underline the difference between the situation we are discussing and the usual idea of a limit, we introduce the concept of *limit in probability*. For example, Bernoulli's theorem can be written in symbolic form:

$$\omega_n \xrightarrow[n\to\infty]{\text{prob}} p \tag{4.3}$$

which we read as: ω_n tends in probability to p as $n \to \infty$.

If in n repeated trials the event A occurs m times, then the ratio m/n will be a particular experimental value of the relative frequency ω_n. For sufficiently large n we can have confidence in the approximate equality

$$\frac{m}{n} \approx p \tag{4.4}$$

with arbitrary previously assigned precision. This means in practice that the value m/n of the relative frequency ω_n possesses the stability of which we spoke at the beginning of the book.

If the probability of the event is unknown, then the relation (4.4) gives an approximate value for this probability from the given experiment. For example, in the population statistics of the nineteenth century it was established that the relative frequency of male births was stable and equal to 0·512. Thus we can conclude that male births have a definite probability near to 0·512.

For a concrete value of n the precision of the approximate equation (4.4) must be estimated: we shall discuss this problem in the following chapter.

§ 15. CHEBYSHEV'S THEOREM

The general law of large numbers for independent random variables was stated and proved by P. L. Chebyshev in 1867.

We consider a sequence of pair-wise independent random variables $\xi_1, \xi_2, \ldots, \xi_n, \ldots$, with arbitrary probability distributions. Let us suppose that all these random variables have well-defined expectations and variances:

$$\mathbf{E}\xi_k = a_k; \quad \mathbf{E}(\xi_k - a_k)^2 = \sigma_k^2 \quad (k = 1, 2, \ldots). \tag{4.5}$$

Let us write

$$\bar{\xi}_n = \frac{\xi_1 + \xi_2 + \cdots + \xi_n}{n} \tag{4.6}$$

for the arithmetic mean of the first n random variables. The mathematical expectation of this arithmetic mean is the arithmetic mean of the mathematical expectations of the variables $\xi_1, \xi_2, \ldots, \xi_n$;

$$\mathbf{E}\bar{\xi}_n = \frac{a_1 + a_2 + \cdots + a_n}{n} = \bar{a}_n \tag{4.7}$$

The variance of the arithmetic mean $\bar{\xi}_n$ is not equal to the arithmetic mean of the variances, but is less by a factor of n:

$$\sigma^2(\bar{\xi}_n) = \frac{1}{n^2}(\sigma_1^2 + \sigma_2^2 + \cdots + \sigma_n^2) = \frac{1}{n}\frac{\sum \sigma_k^2}{n}. \tag{4.8}$$

If all the variances σ_k^2 are bounded by the same number, H,

$$\sigma_k^2 \leqq H \quad (k = 1, 2, \ldots),$$

then the variance of the arithmetic mean tends to zero as $n \to \infty$, since

$$\sigma^2(\bar{\xi}_n) \leqq \frac{H}{n}.$$

It follows from this that as n increases, the value of the arithmetic means $\bar{\xi}_n$ will become less spread out, i.e. more closely grouped near the centre of the distribution.†

Chebyshev's inequality

Chebyshev's inequality gives an estimate of the probability that the deviation of an arbitrary random variable ξ from the centre of its distribution $a = \mathbf{E}\xi$ exceeds a given positive number ε:

$$\mathbf{P}\{|\xi - a| > \varepsilon\} < \frac{\sigma^2(\xi)}{\varepsilon^2}, \tag{4.9}$$

and this probability will be smaller, the smaller the variance $\sigma^2(\xi)$.

We will carry out the proof of Chebyshev's inequality for continuous random variables. From (2.20) we have

$$\mathbf{P}\{|\xi - a| > \varepsilon\} = \int_{|x-a|>\varepsilon} \varphi(x)\, dx,$$

where the integral is taken over the intervals $(-\infty, a - \varepsilon)$ and $(a + \varepsilon, +\infty)$ in which $|x - a| > \varepsilon$. Because the inequality

$$1 < \frac{(x-a)^2}{\varepsilon^2}$$

holds in these intervals, we have also the inequality

$$\varphi(x) \leqq \frac{1}{\varepsilon^2}(x - a)^2\, \varphi(x)$$

and therefore

$$\int_{|x-a|>\varepsilon} \varphi(x)\, dx \leqq \frac{1}{\varepsilon^2} \int_{|x-a|>\varepsilon} (x-a)^2\, \varphi(x)\, dx.$$

† This can be interpreted as saying that by forming the arithmetic means a partial cancellation takes place of random deviations with different signs.

To complete the proof of the inequality (4.9) it is sufficient to remark that

$$\int_{|x-a|>\varepsilon} (x - a)^2 \varphi(x)\, dx \leqq \int_{-\infty}^{\infty} (x - a)^2 \varphi(x)\, dx = \sigma^2(\xi).$$

As an exercise we recommend the reader to carry out the analogous proof of Chebyshev's inequality for discrete variables.

Chebyshev's theorem

We now apply Chebyshev's inequality to the random variable $\bar{\xi}_n$:

$$\mathbf{P}\{|\bar{\xi}_n - \bar{a}_n| > \varepsilon\} < \frac{\sigma^2(\bar{\xi}_n)}{\varepsilon^2} < \frac{H}{n\varepsilon^2}. \tag{4.10}$$

However small we may take the positive number ε, it is always possible to choose a sufficiently large n that the right-hand side of the inequality becomes arbitrarily small. In particular, we can make the right-hand side of (4.10) less than the prescribed bound for "very small" probabilities. Then we will have confidence that in practice the inequality $|\bar{\xi}_n - \bar{a}_n| > \varepsilon$ will not be satisfied, and consequently that the complementary inequality

$$|\bar{\xi}_n - \bar{a}_n| \leqq \varepsilon$$

will hold.

Thus, *for a sufficiently large number of independent random variables we will have confidence that the difference between the arithmetic means of the variables and the centre of their distribution does not exceed in absolute value some prescribed number ε (which may be arbitrarily small).*

From the concentration of the mean $\bar{\xi}_n$ near the centre of its distribution for large n we derive the law of large numbers. We give a precise statement of this law.

CHEBYSHEV'S THEOREM. *If the sequence $\xi_1, \xi_2, \ldots, \xi_n, \ldots$ consists of pair-wise independent random variables with uniformly bounded variances, then the arithmetic means $\bar{\xi}_n$ of the first n variables, $\xi_1, \xi_2, \ldots, \xi_n$ satisfy the relationship*

$$\lim_{n\to\infty} \mathbf{P}\{|\bar{\xi}_n - \mathbf{E}\bar{\xi}_n| > \varepsilon\} = 0 \tag{4.11}$$

for arbitrary positive ε.

A special case of Chebyshev's theorem

If the centre of the distributions of all the random variables $\xi_1, \xi_2, \ldots, \xi_n, \ldots$ is the same number, $\mathbf{E}\xi_k = a\,(k = 1, 2, \ldots)$, then the centre of the distribution of the arithmetic mean $\bar{\xi}_n$ is also equal to a for each n:

$$\mathbf{E}\bar{\xi}_n = \frac{1}{n}(\mathbf{E}\xi_1 + \mathbf{E}\xi_2 + \cdots + \mathbf{E}\xi_n) = a,$$

so that (4.11) becomes

$$\lim_{n\to\infty} \mathbf{P}\{|\bar{\xi}_n - a| > \varepsilon\} = 0. \tag{4.12}$$

This expression can be written symbolically

$$\bar{\xi}_n \xrightarrow[n\to\infty]{\text{prob}} a, \tag{4.13}$$

which is analogous to (4.3).†

The proof of Bernoulli's theorem

If we take the variables ξ_k to be the indicator random variables λ_k, introduced in § 7, then the arithmetic mean will be equal to the relative frequency of the event:

$$\bar{\lambda}_n = \frac{\lambda_1 + \lambda_2 + \cdots + \lambda_n}{n} = \omega_n.$$

In this case $\mathbf{E}\lambda_k = p$, $\sigma^2(\lambda_k) = pq < 1$, so that (4.12) reduces to (4.1) and Bernoulli's theorem is proved.

§ 16. THE STABILITY OF THE SAMPLE MEAN AND THE METHOD OF MOMENTS

We consider first the statistical question of mean values. Suppose we have a collection of N objects, to each of which is attached a number x. Let us select n objects at random (the sample). We

† In 1928, A. Ya. Khinchin proved that for identically distributed, independent random variables (4.12) holds with only one assumption, that the centre a is finite. It is not neccessary that the variances be bounded (they can in fact be infinite).

wish to know whether the mean value of x for the whole collection is near the mean value of the sample.

Chebyshev's theorem gives an answer to this question provided the selection is made according to the scheme of "sampling with replacement" discussed on p. 25. We define for each k, a random variable ξ_k, which takes the value of x observed at the kth drawing. If the sampling is with replacement then each selection is made from the same collection of objects and therefore the random variables will be independent and identically distributed according to (3.2). We showed in § 10 that the mathematical expectation of all these variables coincides with the arithmetic mean of the possible values in the initial collection (the "population"), i.e. with the so-called "population mean", a:

$$\mathbf{E}\xi_k = a \quad (k = 1, 2, \ldots, n).$$

The arithmetic mean

$$\bar{\xi}_n = \frac{\xi_1 + \xi_2 + \cdots + \xi_n}{n}$$

will therefore satisfy the conditions of Chebyshev's theorem in the form (4.12) or (4.13). This states that the mean of the sample values tends in probability to the population mean as the size of the sample increases without limit.

We draw a practical conclusion. The experimental value of each random variable ξ_k is the value which in fact we observe on the kth drawing; the value of $\bar{\xi}_n$ (the arithmetic mean of the observed values) is called the "sample mean", $\bar{x}$. Expression (4.13) can therefore be interpreted as: *For a sufficiently large random sample with replacement we can have confidence that the sample mean will difier by an arbitrarily small amount from the population mean*, i.e. that the approximate equation

$$\bar{x} \approx a \tag{4.14}$$

will hold. We can also infer that the *sample mean possesses stability* in the sense that two sequences of sufficiently long independent Bernoulli trials will have approximately equal sample means. This conclusion agrees well with experiment.

We emphasise that the actual deviation of the sample mean from the population mean depends on the sample size n and not on the ratio of the sample size to the population size.† We remark also that if the size of the population is very large compared with the sample size then it is inessential whether the sampling is carried out with or without replacement. This is very important in practice. The fact is that the population mean is usually unknown and its value is inferred from values of the sample mean. The importance of this is made clear in the following example. We suppose that for some calculations it is necessary to know the mean lifetime of a large batch of electronic valves. In order to obtain the precise value of the mean lifetime for the whole batch we would have to test each lamp: but then there would be no lamps left in the population whose mean we had obtained. In practice we obtain a mean value of this kind by sampling. Finally, we need an estimate of the precision of this method of obtaining the population mean from the sample mean: we will discuss this question in Chapter VI.

On the method of moments

The approximate equation (4.14) can be given another interpretation. Let ξ be a random variable with finite expectation $\mathbf{E}\xi = a$. If we carry out independent trials, ξ will take definite values $x_1, x_2, \ldots, x_n$. These experimental values of the variable ξ can be considered as values of *distinct* random variables $\xi_1, \xi_2, \ldots, \xi_n$, with the same probability distribution as the variable ξ, and they will, moreover, be independent because the trials are independent. On this interpretation the arithmetic mean of the experimental data

$$\bar{x} = \frac{x_1 + x_2 + \cdots + x_n}{n}$$

is an experimental value of the variable $\bar{\xi}_n$ for which Chebyshev's theorem (4.12) holds. For sufficiently large n we can therefore expect that the approximate equation

$$\bar{x} \approx \mathbf{E}\xi = a \tag{4.15}$$

will hold with any prescribed accuracy.

Thus, *an approximate value of the mathematical expectation of*

† For example, a 1% sample from a population 1,000,000 gives more precise information about the population mean than a 2% sample from a population of 1000 (for the same value of σ).

a random variable can be obtained by taking the artihmetic mean of values obtained experimentally.

This procedure enables us to find approximations for other moments of the distribution, which are defined in the same way as the mathematical expectation for some variables. For example, we obtain the following approximate relation for the variance:

$$\sigma^2(\xi) = \mathbf{E}(\xi - a)^2 \approx \frac{\sum (x_k - a)^2}{n}, \tag{4.16}$$

where the sum is taken over all the experimental values $x_1, x_2, \ldots, x_n$.

In fact, the value $\sum (x_k - a)^2/n$ can be considered as the arithmetic mean of n identically distributed, independent random variables $(\xi_k - a)^2$ with mathematical expectation

$$\mathbf{E}(\xi_k - a)^2 = \mathbf{E}(\xi - a)^2 = \sigma^2(\xi) \quad (k = 1, 2, \ldots, n).$$

Therefore

$$\mathbf{P}\left\{\left|\frac{\sum (\xi_k - a)^2}{n} - \sigma^2(\xi)\right| > \varepsilon\right\} \to 0 \quad \text{as} \quad n \to \infty.$$

In (4.16) the value of the centre of the distribution, a, was used. This number is usually unknown and it is natural to try replacing it by the approximate value $\bar{x}$. However, while the resulting approximate equation

$$\sigma^2(\xi) \approx \frac{\sum (x_k - \bar{x})^2}{n} \tag{4.17}$$

is valid, it is not the precise parallel of (4.15) and (4.16). The fact is that although the values $\sum (x_k - \bar{x})^2/n$ can be considered as particular values of the arithmetic mean of n random variables $(\xi_k - \bar{\xi}_n)^2$, the mathematical expectation of these variables is not equal to $\sigma^2(\xi)$. We calculate the mathematical expectation directly:

$$\begin{aligned}
\mathbf{E}(\xi_1 - \bar{\xi}_n)^2 &= \mathbf{E}\left[(\xi_1 - a) - \left(\frac{\xi_1 + \xi_2 + \cdots + \xi_n}{n} - a\right)\right]^2 \\
&= \mathbf{E}\left[(\xi_1 - a)\left(1 - \frac{1}{n}\right)\right. \\
&\quad \left. - \left(\frac{\xi_2 - a}{n}\right) - \cdots - \left(\frac{\xi_n - a}{n}\right)\right]^2 \\
&= \sigma^2(\xi_1)\left(1 - \frac{1}{n}\right)^2 + \frac{\sigma^2(\xi_2)}{n^2} + \cdots + \frac{\sigma^2(\xi_n)}{n^2};
\end{aligned}$$

and similarly for the other k. Hence

$$\mathbf{E}(\xi_k - \bar{\xi}_n)^2 = \frac{n-1}{n}\sigma^2(\xi)\,.\dagger$$

Using the linearity of the mathematical expectation it follows that the variables

$$\frac{n}{n-1}(\xi_k - \bar{\xi}_n)^2$$

will have the expectation $\sigma^2(\xi)$. Although they are not independent, the law of large numbers holds for them‡ and the value of their arithmetic mean

$$\frac{\sum \frac{n}{n-1}(\xi_k - \bar{\xi}_n)^2}{n} = \frac{\sum (\xi_k - \bar{\xi}_n)^2}{n-1}$$

will be arbitrarily close to $\sigma^2(\xi)$ for sufficiently large n.

Thus, we can correct (4.17) by introducing the factor $n/(n-1)$ into the right-hand side. For the calculation of the variance from experimental data the correct approximate formula is

$$\sigma^2(\xi) \approx \frac{\sum (x_k - \bar{x})^2}{n-1}. \tag{4.18}$$

The variable on the right-hand side of this relation is called the *sample variance* and is denoted by s_n^2. We should remark that for large n the number $n - 1$ differs relatively little from n so that formulae (4.17) and (4.18) give practically the same results. However, for small values of n the difference between them is quite noticeable.

Finally, all the approximate equations which we have discussed require an estimate of error. We will give some estimates in Chapter VI.

If the form of a distribution is known, then the parameters of the distribution can be found by using the approximate values of the moments. We illustrate this procedure with examples.

† The cause of this is related to the linear dependence between the variables $\xi_1, \xi_2, \ldots, \xi_n$, and their mean $\bar{\xi}_n = (1/n)(\xi_1 + \xi_2 + \cdots + \xi_n)$.

‡ We omit the proof of this.

(1) In the general normal distribution with density

$$\varphi(x) = \frac{1}{\sigma\sqrt{(2\pi)}} \exp\left(-\frac{(x-a)^2}{2\sigma^2}\right)$$

the parameters a and σ^2 are respectively the centre and the variance (see § 12). They can be estimated immediately from experimental data using expressions (4.15) and (4.18).

(2) The unknown parameters in the uniform distribution are the end points of the interval (α_1, α_2).

As we saw in § 12, the moments of this distribution are expressed in terms of these parameters by the relations

$$a = \mathbf{E}\xi = \frac{\alpha_1 + \alpha_2}{2}; \qquad \sigma = \sigma(\xi) = \frac{\alpha_2 - \alpha_1}{2\sqrt{3}}.$$

Consequently we obtain

$$\alpha_1 = a - \sigma\sqrt{3}; \quad \alpha_2 = a + \sigma\sqrt{3},$$

where a and σ are given by (4.15) and (4.18).

(3) For the Pearson distribution (2.27) the parameters α and β are connected with the centre and variance by the formulae

$$a = \mathbf{E}\xi = \frac{\alpha}{\beta}; \quad \sigma^2 = \sigma^2(\xi) = \frac{\alpha}{\beta^2}$$

(see Ex. (10), Chapter III). Whence

$$\alpha = \frac{a^2}{\sigma^2}; \quad \beta = \frac{a}{\sigma^2},$$

where a and σ^2 are given by (4.15) and (4.18).

If it is known that the probability density function depends on l unknown parameters $\alpha_1, \alpha_2, \ldots, \alpha_l$, then we can express the first l moments of the distribution in terms of them and thus derive l equations from which (in principle) we can find the l parameters. We can find these moments from experimental data by the method indicated above. However, the higher the order of the moments of the distribution the more experimental data will be needed in order to estimate the moments with any reasonable precision. Consequently in practice we often limit ourselves to distributions with only two unknown parameters which we can find with the help of the first two moments of the distribution.

EXERCISES

(1) Observations were made at a telephone exchange throughout a period of one hour on the number of incorrect junctions made in a minute: the numbers were as follows.

3	1	3	1	4	2
2	4	0	3	0	2
2	0	2	1	4	3
3	1	4	2	2	1
1	2	1	0	3	4
1	3	2	7	2	0
0	1	3	3	1	2
4	2	0	2	3	1
2	5	1	1	0	1
1	2	2	1	1	5

Find the centre and variance of this distribution and verify that the basic property of the Poisson distribution $\mathbf{E}\xi = \sigma^2 = a$ is satisfied. Find the corresponding Poisson distribution. Compare the frequency distribution table of the data with the corresponding table of the Poisson distribution (Table IV).

(2) The measurements of 100 machined parts gave the following deviations from a nominal value:

−2	2	1	2	−1	−2	3	1	−1	0
0	−1	3	1	2	−3	1	0	1	1
0	1	−1	1	0	2	2	1	0	−1
1	1	4	−1	1	1	−1	0	2	−2
2	0	−2	0	0	−1	1	4	−2	1
−3	0	0	1	4	0	−2	2	1	2
−1	1	0	−1	0	3	1	−2	3	−1
1	2	2	0	−2	1	0	−1	0	3
3	−2	−1	−2	1	0	0	−3	1	0
2	1	0	3	−1	2	1	0	−1	0

Find the centre and variance of the distribution and construct the corresponding normal distribution. Compare the table of the cumulative relative frequency of the data with the corresponding table of the cumulative normal distribution function (Table I).

(3) Fit a Pearson distribution (2.27) to the following data

Value x	0	1	2	3	4	5	6	
Frequency m	1	33	41	18	5	1	1	100

Verify that the basic property $C_s = 2C_v$ of the Pearson distribution is satisfied (see Chapter III, Exercise 10). Compare the cumulative frequency table of the data with the corresponding cumulative distribution function.

(4) Chebyshev's inequality is very weak. Consider, for example, the distribution (2.2). Observe directly that $\mathbf{P}\{|\xi - 3{\cdot}5| > 3\} = 0$, and show that Chebyshev's inequality gives an upper bound of approximately $\frac{1}{3}$ (use Chapter II, Exercise 3).

(5) If ω_n is the relative frequency of the occurrence of heads in the repeated tossing of a fair coin ($p = q = \frac{1}{2}$), use Chebyshev's inequality to find an upper bound to the number, n, of trials required so that $\mathbf{P}\{|\omega_n - \frac{1}{2}| > 0{\cdot}01\} \leqq 0{\cdot}01$.

(6) Let $\xi_1, \xi_2, \ldots, \xi_n, \ldots$ be a sequence of random variables with $\mathbf{E}\xi_n = 0$, $\sigma^2(\xi_n) = 1$ such that ξ_n depends only on ξ_{n-1} and ξ_{n+1} and not on the other ξ_m. Show that (4.11) still holds.

(7) Let $\xi_1, \xi_2, \ldots, \xi_n$ be a sequence of dependent random variables with $\mathbf{E}\xi_n = 0, \sigma^2(\xi_n) = 1$. If $\varrho_{nm} = \mathbf{E}\xi_n\xi_m \to 0$ uniformly as $|n - m| \to \infty$, show that (4.11) still holds. (Bernstein's Theorem.)

(8) *Khinchin condition* (footnote, p. 76). By Chapter III, Exercise 20, the distribution given by (2.44) does not give rise to a relation of the form of (4.12). Show that, for α satisfying $1 < \alpha \leqq 2$, the following distribution has a finite centre but infinite variance:

$$F(x) = \begin{cases} \frac{1}{2}|x|^{-\alpha} & \text{for } x \leqq -1 \\ \frac{1}{2} & \text{for } -1 \leqq x \leqq 1 \\ 1 - \frac{1}{2}x^{-\alpha} & \text{for } 1 \leqq x. \end{cases}$$

Random variables with this distribution will therefore satisfy (4.12).†

† For a proof of Khinchin Theorem when the random variables are discrete see W. Feller, *An Introduction to Probability Theory and its Applications*, 2nd edition (1957).

CHAPTER V

LIMIT THEOREMS AND ESTIMATES OF THE MEAN

IN ESTIMATING relative frequencies and some other mean values it is of decisive importance that the distributions of these variables tend to the normal distribution. For the proof of these limit theorems the well-known Russian mathematician A. M. Lyapunov exploited the very powerful method of characteristic functions which enabled him to prove the so-called Central Limit Theorem (see § 19). Before discussing the limit theorems, however, we give some necessary information on characteristic functions.

§ 17. THE CHARACTERISTIC FUNCTION

The characteristic function $f(u)$ of the random variable ξ is defined to be the mathematical expectation of the variable $e^{iu\xi}$:

$$f(u) = \mathbf{E}e^{iu\xi}, \tag{5.1}$$

where u is a real parameter.

For a discrete random variable

$$f(u) = \sum e^{iux_k} p_k, \tag{5.2}$$

where p_k is the probability of the value x_k and the sum is taken over the range of all possible values of ξ. For a continuous random variable

$$f(u) = \int_{-\infty}^{\infty} e^{iux} \varphi(x)\, dx, \tag{5.3}$$

where $\varphi(x)$ is the probability density function of the variable ξ. The integral (5.3) always converges, and in fact converges absolutely since $|e^{iux}\varphi(x)| = \varphi(x)$, and the integral of $\varphi(x)$ converges: thus

$$|f(u)| \leq \int_{-\infty}^{\infty} \varphi(x)\, dx = 1.$$

Fundamental properties of characteristic functions

(1) A characteristic function uniquely determines a distribution function.† In other words, if two random variables have identical characteristic functions, then they have the same distribution function.

(2) We suppose that $f(u)$ is the characteristic function of a *continuous* random variable ξ, and that $f_n(u)$ form a sequence of characteristic functions of arbitrary random variables ξ_n $(n = 1, 2, 3, \ldots)$. If $f(u)$ is the limit as $n \to \infty$ of the sequence $f_n(u)$, then the cumulative distribution function $F(x) = \mathbf{P}\{\xi < x\}$ is the limit of the sequence of cumulative distribution functions $F_n(x) = \mathbf{P}\{\xi_n < x\}$. Thus, if $\lim_{n\to\infty} f_n(u) = f(u)$, it follows that $\lim_{n\to\infty} F_n(x) = F(x)$ for all n.‡

This property is important since in many cases it is easier to carry out the passage to the limit in the sequence of characteristic functions than in the sequence of distribution functions. The proofs of limit theorems in these cases are therefore much shorter and simpler using characteristic functions.

We will accept the above properties without proof.

(3) The characteristic function of a sum of independent random variables is equal to the product of the characteristic functions.

We shall establish this property for two independent random variables ξ and η with characteristie functions $f_\xi(u)$ and $f_\eta(u)$. Let $f_{\xi+\eta}(u)$ be the characteristic function of the sum $\xi + \eta$. Because the random variables $e^{iu\xi}$ and $e^{iu\eta}$ will also be independent, we can apply the multiplication theorem for mathematical expectations to obtain

$$f_{\xi+\eta}(u) = \mathbf{E}e^{iu(\xi+\eta)} = \mathbf{E}e^{iu\xi}e^{iu\eta} = \mathbf{E}e^{iu\xi}\,\mathbf{E}e^{iu\eta},$$

so that

$$f_{\xi+\eta}(u) = f_\xi(u)\, f_\eta(u). \tag{5.4}$$

Thus it is simpler to find the characteristic function of a sum of independent random variables than it is to find the corresponding distribution function (which involves the convolution of probability density functions: see § 9).

(4) If the random variable ξ is transformed into the random variable $\eta = A + B\xi$ (i.e., linearly), then the characteristic function is transformed according to the relation

$$f_\eta(u) = e^{iAu} f_\xi(Bu). \tag{5.5}$$

We verify this immediately:

$$\mathbf{E}e^{iu\eta} = \mathbf{E}e^{iu(A+B\xi)} = e^{iAu}\,\mathbf{E}e^{iuB\xi}.$$

† It is possible to establish a relation expressing a distribution function in terms of its characteristic function: see B. V. Gnedenko, *Theory of Probability*, New York, 1962, ch. 7. In the same reference will be found a proof of our statement. For the reader acquainted with Fourier integrals, we remark that (5.3) is the Fourier transform of the density $\varphi(x)$.

‡ See Gnedenko for a more general theorem of this kind.

EXAMPLES OF CHARACTERISTIC FUNCTIONS

(1) The characteristic function of the random variable λ $\begin{array}{c|c} 1 & 0 \\ \hline p & q \end{array}$ is easily calculated from (5.2):

$$f_\lambda(u) = e^{iu1}p + e^{iu0}q = pe^{iu} + q.$$

(2) The frequency of an event in n trials is the sum

$$\mu_n = \lambda_1 + \lambda_2 + \cdots + \lambda_n,$$

where the λ_k are independent and identically distributed with

λ_k	1	0
	p	q

($k = 1, 2, \ldots, n$) (see § 7). The characteristic function of μ_n follows from Property 3:

$$f_{\mu_n}(u) = f_{\lambda_1}(u) f_{\lambda_2}(u) \ldots f_{\lambda_n}(u) = (pe^{iu} + q)^n. \tag{5.6}$$

(3) The relative frequency of an event is $\omega_n = \mu_n/n$; we derive its characteristic function from (5.5) and (5.6):

$$f_{\omega_n}(u) = f_{\mu_n}(u/n) = (pe^{i(u/n)} + q)^n.$$

(4) A random variable which is uniformly distributed on the interval $(-a, a)$ has the density function

$$\varphi(x) = \begin{cases} \dfrac{1}{2a} & (-a < x < a) \\ 0 & (|x| > a). \end{cases}$$

Its characteristic function is obtained using (5.3),

$$f(u) = \int_{-a}^{a} e^{iux} \frac{1}{2a}\, dx = \frac{e^{iua} - e^{-iua}}{2aiu} = \frac{\sin au}{au}. \tag{5.7}$$

(5) If ξ_0 is a random variable with the standard normal distribution (2.22), then its characteristic function is†

$$f_0(u) = \frac{1}{\sqrt{(2\pi)}} \int_{-\infty}^{\infty} e^{iux}\, e^{-x^2/2}\, dx = e^{-u^2/2}. \tag{5.8}$$

A random variable ξ with the general normal distribution (2.26) is linearly related to ξ_0 by the expression $\xi = a + \sigma\xi_0$. We calculate its characteristic function from (5.8) and (5.5):

$$f(u) = e^{iau} f_0(\sigma u) = e^{iau}\, e^{-\sigma^2 u^2/2}. \tag{5.9}$$

† The evaluation of this integral is assumed.

It follows from this that if the mutually independent random variables ξ_1, $\xi_2, \ldots, \xi_n$ have general normal distributions with centres $a_1, a_2, \ldots, a_n$, and variances $\sigma_1^2, \sigma_2^2, \ldots, \sigma_n^2$, then their sum is normally distributed with centre $a = a_1 + a_2 + \cdots + a_n$ and variance $\sigma_2 = \sigma_1^2 + \sigma_2^2 + \cdots + \sigma_n^2$. In fact, if $f_k(u) = e^{ia_k u}\, e^{-\sigma_k^2 u^2/2}$ $(k = 1, 2, \ldots, n)$, then $f(u) = f_1(u)\, f_2(u) \ldots f_n(u) = e^{iau}\, e^{-\sigma^2 u^2/2}$.

The connection between the characteristic function and the moments of a distribution

Since the characteristic function $f(u)$ uniquely determines the probability distribution of the random variable ξ, it follows that we can express all the moments in terms of the characteristic function. To derive these expressions we differentiate the equation

$$f(u) = \mathbf{E} e^{iu\xi}$$

formally with respect to u (under the expectation sign, i.e. under the summation or integral sign). We obtain from this

$$f'(u) = \mathbf{E} i\xi e^{iu\xi}$$

$$f''(u) = \mathbf{E}(i\xi)^2 e^{iu\xi}$$

$$\cdots\cdots\cdots\cdots\cdots$$

$$f^{(k)}(u) = \mathbf{E}(i\xi)^k e^{iu\xi}.$$

It can be shown that this differentiation is legitimate if the random variable possesses moments up to and including the kth order. If we now put $u = 0$ in these expressions we obtain relations between the derivative at the origin of the characteristic function and the primitive moments:

$$f(0) = \mathbf{E}\,1 = 1,$$

$$f'(0) = i\mathbf{E}\xi,$$

$$f''(0) = -\mathbf{E}\xi^2,$$

$$\cdots\cdots\cdots\cdots\cdots$$

$$f^{(k)}(0) = i^k \mathbf{E}\xi^k.$$

The logarithm of the characteristic function and its derivatives also occur in applications. Let $\psi(u) = \ln f(u)$. The number $i^k \psi^{(k)}(0)$ is called the *semi-invariant*† of order k of the random variable ξ. It is easy to verify that

$$i\psi'(0) = -\mathbf{E}\xi; \quad i^2\psi''(0) = \sigma^2(\xi).$$

The semi-invariants play an important role in the study of sums of independent random variables, because if we add independent random variables then their semi-invariants are added also.

† *Translator's note:* The Anglo-American usage 'cumulant' is extensive.

§ 18. THE LIMIT THEOREM OF DE MOIVRE–LAPLACE. ESTIMATION OF THE RELATIVE FREQUENCY

In this section we shall study the limiting distribution of the relative frequency of an event as the number of trials increases without limit. As we have seen (see § 7), the relative frequency ω_n (for a scheme of random sampling with replacement) has the binomial distribution:

$$\mathbf{P}\left\{\omega_n = \frac{m}{n}\right\} = \binom{n}{m} p^m q^{n-m} \quad (m = 0, 1, 2, \ldots, n), \quad (5.10)$$

where n is the number of trials, and p is the probability of occurrence of the event at each trial. If the number of trials is large, then the calculation of probabilities from (5.10) becomes very difficult. The nature of the difficulty becomes clear if we recall that in practical applications we are not so much interested in the probability of an individual event ($\omega_n = m/n$), but in the probability of a compound event ($|\omega_n - p| < \varepsilon$) involving the deviation of the relative frequency from its limiting probability (see § 14). The probability of this compound event is equal to the sum $\sum \binom{n}{m} p^m q^{n-m}$ taken over all values of m for which $|(m/n) - p| < \varepsilon$, i.e. for those values of m lying between $np - n\varepsilon$ and $np + n\varepsilon$. The difficulties arising in this calculation will be clear from an example. Suppose we are interested in the probability that in 10,000 trials the deviation of the relative frequency of an event from its probability 0·2 does not exceed $\varepsilon = 0{\cdot}01$. Here, $n = 10{,}000$, $np = 2000$, $q = 0{\cdot}8$, and to calculate accurately the probability we are interested in requires that the evaluate the sum of more than two hundred terms of the form

$$\frac{10{,}000!}{m!\,(10{,}000 - m)!}(0{\cdot}2)^m\,(0{\cdot}8)^{10{,}000-m}$$

from $m = np - n\varepsilon = 1900$ to $np + n\varepsilon = 2100$. The direct evaluation of such a sum with any satisfactory accuracy requires a vast expenditure of work. The question therefore arose a long time ago of approximating for large values of n to the exact binomial distribution of the variable ω_n with some continuous distributions.† The problem was successfully solved by de Moivre (in 1730) for the special case $p = q = \frac{1}{2}$, and then by Laplace (in 1783) for an arbitrary value of p satisfying $0 < p < 1$. They found that the binomial distribution tends to a limiting distribution as $n \to \infty$, and that this limiting distribution is normal.

For convenience in formulating these theorems we will first of all "normalise" the relative frequency ω_n.

The random variable

$$\xi_0 = \frac{\xi - \mathbf{E}\xi}{\sigma(\xi)}$$

is called the normalised form of the variable ξ. This linear transformation involves a shift of centre and a change in the scale of measurement. Thus $\mathbf{E}\xi_0 = 0$, and $\sigma(\xi_0) = 1$.

We shall write τ_n for the normalised form of the relative frequency:

$$\tau_n = \frac{\omega_n - \mathbf{E}\omega_n}{\sigma(\omega_n)} = \frac{\omega_n - p}{\sqrt{\left(\frac{pq}{n}\right)}}. \tag{5.11}$$

The theorem of de Moivre–Laplace

As the number of trials increases without limit the distribution of the normalised relative frequency tends to the standardised normal distribution: i.e.,

$$\lim_{n\to\infty} \mathbf{P}\{|\tau_n| < t\} = \Phi(t), \tag{5.12}$$

where $\Phi(t)$ *is the normal probability integral* (2.23).

This theorem is a special case of a more general theorem which we shall discuss in the next section. We shall examine here the character of the de Moivre–Laplace theorem by means of some examples.

† The calculation of the probabilities of inequalities reduces to the evaluation of integrals when we have continuous distributions, and this is usually much simpler than the evaluation of sums for discrete distributions.

The application of the de Moivre–Laplace theorem to estimates of relative frequencies

The theorem of de Moivre–Laplace enables us to estimate the probability of the event ($|\omega_n - p| < \varepsilon$) for sufficiently large n (and for p not too near to 0 or 1). We shall take a large enough n so that

$$\mathbf{P}\{|\tau_n| < t\} \approx \Phi(t) \tag{5.13}$$

with sufficient accuracy for our purposes. Then from the equivalent inequalities

$$|\omega_n - p| < \varepsilon \quad \text{and} \quad |\tau_n| = \left|\frac{\omega_n - p}{\sqrt{(pq/n)}}\right| < \frac{\varepsilon\sqrt{n}}{\sqrt{(pq)}}$$

it follows that the probability of the event ($|\omega_n - p| < \varepsilon$) which is of interest to us is approximately equal to the normal probability integral

$$\mathbf{P}\{|\omega_n - p| < \varepsilon\} \approx \Phi(t), \quad \text{where} \quad t = \frac{\varepsilon\sqrt{n}}{\sqrt{(pq)}}. \tag{5.14}$$

EXAMPLES. Let us return to the numerical example at the beginning of this section: we had $p = 0{\cdot}2$; $q = 0{\cdot}8$; $n = 10{,}000$; and $\varepsilon = 0{\cdot}01$. We calculate the probability of the event ($|\omega_n - p| < \varepsilon$) using (5.14):

$$t = \frac{0{\cdot}01\sqrt{10{,}000}}{\sqrt{(0{\cdot}2 \times 0{\cdot}8)}} = 2{\cdot}5;$$

$$\mathbf{P}\{|\omega_n - 0{\cdot}2| < 0{\cdot}01\} \approx \Phi(2{\cdot}5) = 0{\cdot}988.\dagger$$

† In this case the approximate formula is correct to the third figure. This can be accounted for by the large value of n (10,000). Precise estimates of the error involved in using (5.14) can be found in the work of S. N. Bernstein, W. Feller, and others. We remark that if the number n is only of the order of several hundred (with np and nq still significantly greater than 1), then in place of (5.14) one should use the somewhat more accurate formula

$$\mathbf{P}\left\{|\omega_n - p| \leqq \frac{k}{n}\right\} \approx \Phi\left(\frac{k+\frac{1}{2}}{\sqrt{(npq)}}\right) \quad (k \text{ an integer}).$$

We exhibit the accuracy of this formula by means of examples.

(1) $p = \frac{1}{2}$; $n = 200$; $k = 5$;

$$\mathbf{P}\left\{|\omega_n - \tfrac{1}{2}| \leqq \frac{5}{200}\right\} = 0{\cdot}56325;$$

[*continued on page* 91]

If we consider the probability $P = 0{\cdot}988$ to be sufficiently near to one, then we can be confident that in practice $|\omega_n - 0{\cdot}2| < 0{\cdot}01$, i.e., that in 10,000 trials (sampling with replacement) the relative frequency of the event will differ from its probability, $p = 0{\cdot}2$, by less than 0·01. The probability 0·988 is called the "*reliability*" *of the estimate*

$$|\omega_n - 0{\cdot}2| < 0{\cdot}01 .$$

In practice the reliability of an estimate is given beforehand in terms of the prescribed bound for 'very small probabilities' (see § 13). For example, if we have agreed to neglect the possibility of occurrence of an event with probability 0·001, then the prescribed reliability is $P = 1 - 0{\cdot}001 = 0{\cdot}999$.

For a given reliability we can find a value $t = t(P)$ from the equation

$$\Phi(t) = P,$$

using the tables of the normal probability integral; e.g., $P = 0{\cdot}999$ gives $t = 3{\cdot}29$. Then an interval with reliability 0·999 takes the form

$$|\omega_n - p| < \varepsilon, \quad \text{where} \quad \varepsilon = t(P)\sqrt{(pq/n)}. \tag{5.15}$$

The inequality (5.15) means that the relative frequency ω_n must lie in the interval $(p - \varepsilon, p + \varepsilon)$, with reliability $P = 0{\cdot}999$ (where $\varepsilon = t(P)\sqrt{(pq/n)}$. Such an interval is called a '*reliability interval*'.†

$$\frac{1}{\sqrt{(npq)}} = \frac{1}{\sqrt{50}} = 0{\cdot}14142; \quad \Phi\left(\frac{5{\cdot}5}{\sqrt{(npq)}}\right) = 0{\cdot}56331 .$$

(2) The accuracy is less for a non-symmetric interval. Thus if $p = 0{\cdot}1$; $n = 500$; $k = 5$; $\mathbf{P}\{0{\cdot}1 \leqq \omega_n \leqq 0{\cdot}11\} = 0{\cdot}3176$,

$$\frac{1}{\sqrt{(npq)}} = \frac{1}{\sqrt{45}} = 0{\cdot}1491; \quad \frac{1}{2}\Phi\left(\frac{5{\cdot}5}{\sqrt{(npq)}}\right) - \frac{1}{2}\Phi\left(\frac{-0{\cdot}5}{\sqrt{(npq)}}\right) = 0{\cdot}3235,$$

where we have made use of (2.24).

† *Translator's note:* The author used the term 'confidence interval'. In Anglo-American usage this term is applied to statements of the form (6.7) and (6.8) which the author has called 'classical estimates'. I have introduced the term 'reliability interval': no name is usually given to it by Anglo-American writers on statistics.

In the above example the reliability interval of the relative frequency with reliability 0·999 will be

$$|\omega_n - 0{\cdot}2| < 3{\cdot}29 \sqrt{\left(\frac{0{\cdot}2 \times 0{\cdot}8}{10{,}000}\right)} = 0{\cdot}0132.$$

This means that with reliability 0·999 we can expect that the relative frequency ω_n will lie in the reliability interval (0·1868, 0·2132). The corresponding interval for the frequency $\mu_n = n\omega_n$ will be n times larger: in this case $n = 10{,}000$ and the reliability interval is $1868 < \mu_n < 2132$.

The practical value of reliability intervals is not only to be found in the possibility of specifying in advance the limits for the frequency (or limits for the relative frequency). If as a result of actually carrying out a sequence of trials we obtain a value of the frequency lying outside the reliability interval, then we must doubt the correctness of our assumption that the probability of the event is p (in the above example $p = 0{\cdot}2$). This statement of the problem is very useful in the regulation of a mass production process. We illustrate this by the following example. Let us suppose that the automatic machining of components is so regulated that the proportion of defective items does not exceed 1%. In order to check that an increase in this proportion does not take place during production (i.e. that the process does not change) we can introduce a sampling method of control. In a sample of n items† the frequency of defective items $\mu_n = n\omega_n$ must lie in the reliability interval

$$n(p - \varepsilon) < \mu_n < n(p + \varepsilon),$$

i.e. μ_n must not exceed the number $np + n\varepsilon$, where $\varepsilon = t(P)\sqrt{(pq/n)}$, P is the reliability, and p is the assumed probability of finding a defective item. If, for example, $p = 0{\cdot}01$ (1%), $n = 1000$, and $P = 0{\cdot}999$, then $np = 10$, $t = 3{\cdot}29$, $n\varepsilon = 3{\cdot}29 \sqrt{(1000 \times 0{\cdot}01 \times 0{\cdot}99)} = 10{\cdot}4$; $np + n\varepsilon = 20{\cdot}4$. This means that the number of defective items must not exceed twenty. However, if, in a sample, the number of defective items turns out to be greater than twenty, then we

† The sample should be selected randomly and with replacement. However, if the size of the sample is very small compared with the size of the entire batch of components, then the exhibited formulae are sufficiently accurate even for sampling without replacement.

must conclude that the process has changed and the proportion of defective items is greater than the permitted 1%.

§ 19. RELIABILITY INTERVALS FOR MEANS. THE CENTRAL LIMIT THEOREM OF LYAPUNOV

The normality of the limiting distribution in the de Moivre–Laplace theorem is not the result of some special property of the binomial distribution but follows rather from the fact that the relative frequency ω_n is the arithmetic mean of independent random variables $\lambda_1, \lambda_2, \ldots, \lambda_n$. We proceed to an immediate generalisation for arithmetic means of arbitrary sequences of independent identically distributed random variables (we shall assume that the centre and variance of the distribution are finite).

Let $\xi_1, \xi_2, \ldots, \xi_n, \ldots$ be a sequence of mutually independent random variables, each with the same distribution: we suppose $\mathbf{E}\xi_k = a$ and $\mathbf{E}(\xi_k - a)^2 = \sigma^2$ $(k = 1, 2, 3, \ldots)$. We shall denote the arithmetic mean of the first n variables by

$$\bar{\xi}_n = \frac{\xi_1 + \xi_2 + \cdots + \xi_n}{n};$$

for convenience in formulating the theorem we introduce the normalised means (or normalised sums):

$$\tau_n = \frac{\bar{\xi}_n - \mathbf{E}\bar{\xi}_n}{\sigma(\bar{\xi}_n)} = \frac{(\xi_1 + \xi_2 + \cdots + \xi_n) - \mathbf{E}(\xi_1 + \xi_2 + \cdots + \xi_n)}{\sigma(\xi_1 + \xi_2 + \cdots + \xi_n)}. \tag{5.16}$$

Because the random variables ξ_k are independent we have $\mathbf{E}\bar{\xi}_n = a$; $\sigma(\bar{\xi}_n) = \sigma/\sqrt{n}$ (see Chapter III). Hence

$$\tau_n = \frac{\bar{\xi}_n - a}{\sigma/\sqrt{n}} = \frac{(\xi_1 + \xi_2 + \cdots + \xi_n) - na}{\sigma\sqrt{n}}.$$

THEOREM. The limiting distribution (as $n \to \infty$) of the normalised means (5.16) is the standard normal distribution, i.e.

$$\lim_{n\to\infty} \mathbf{P}\{|\tau_n| < t\} = \Phi(t), \tag{5.17}$$

where $\Phi(t)$ is the normal probability integral (2.23).

The proof of this theorem makes use of the method of characteristic functions. We shall confine our attention to continuous random variables. Let $\varphi(x)$ be the probability density function of the normalised random variable $\xi_k^{(0)} = (\xi_k - a)/\sigma$ (the same for all k, by assumption). Then

$$\int \varphi(x)\,dx = 1,$$

$$\int x\varphi(x)\,dx = \mathbf{E}\xi_k^{(0)} = \frac{\mathbf{E}(\xi_k - a)}{\sigma} = 0,$$

$$\int x^2\varphi(x)\,dx = \mathbf{E}[\xi_k^{(0)}]^2 = \frac{\mathbf{E}(\xi_k - a)^2}{\sigma^2} = 1.$$

We shall denote the corresponding characteristic function by $f(u)$:

$$f(u) = \int e^{iux}\varphi(x)\,dx.$$

Let us consider now the sequence of functions $f_n(u)$ which are the characterstic functions of the random variables τ_n. In order to express $f_n(u)$ in terms of $f(u)$ we must first express τ_n in terms of the $\xi_k^{(0)}$:

$$\tau_n = \frac{1}{\sqrt{n}} \frac{(\xi_1 - a) + (\xi_2 - a) + \cdots + (\xi_n - a)}{\sigma}$$

$$= \frac{1}{\sqrt{n}}(\xi_1^{(0)} + \xi_2^{(0)} + \cdots + \xi_n^{(0)}).$$

Since the $\xi_k^{(0)}$ are independent we make use of properties (3) and (4) of characteristic functions (see p. 85) to derive

$$f_n(u) = \left[f\left(\frac{u}{\sqrt{n}}\right)\right]^n,$$

or

$$f_n(u) = \left[\int e^{i\frac{u}{\sqrt{n}}x}\varphi(x)\,dx\right]^n.$$

Expanding the function under the integral sign in a series of powers of $1/\sqrt{n}$ (and stopping with the term in $1/n$) we obtain

$$\int e^{i\frac{u}{\sqrt{n}}x}\varphi(x)\,dx = \int\left(1 + i\frac{u}{\sqrt{n}}x - \frac{u^2}{2n}x^2 + \cdots\right)\varphi(x)\,dx$$

$$= \int \varphi(x)\,dx + i\frac{u}{\sqrt{n}}\int x\varphi(x)\,dx - \frac{u^2}{2n}\int x^2\varphi(x)\,dx + \cdots,$$

so that

$$\int e^{i\frac{u}{\sqrt{n}}x}\varphi(x)\,dx = 1 - \frac{u^2}{2n} + \frac{\alpha_n}{n},$$

where $\alpha_n \to 0$ as $n \to \infty$.†

It is now easy to find the limit of the sequence

$$f_n(u) = \left(1 - \frac{u^2}{2n} + \frac{\alpha_n}{n}\right)^n.$$

It is a well-known theorem in calculus that

$$\lim_{n\to\infty} f_n(u) = e^{-u^2/2}.$$

Thus the sequence of characteristic functions $f_n(u)$ converges to the characteristic $f_0(u) = e^{-u^2/2}$ of the standard normal distribution (see § 17). It follows from this that the sequence of distribution functions of the normalised means τ_n will converge to the distribution function of the standard normal distribution, and this statement is equivalent to (5.17).

Having proved this theorem we can now use it to obtain reliability intervals for mean values, i.e., to find a number ε such that the inequality $|\bar{\xi}_n - a| < \varepsilon$ holds with reliability P. With this in mind we replace the inequality $|\bar{\xi}_n - a| < \varepsilon$ by the equivalent inequality

$$\left|\frac{\bar{\xi}_n - a}{\sigma/\sqrt{n}}\right| < \frac{\varepsilon\sqrt{n}}{\sigma}, \quad \text{i.e.} \quad |\tau_n| < t = \frac{\varepsilon\sqrt{n}}{\sigma}.$$

These inequalities are equivalent in the sense that the events which they determine have equal probabilities. We have proved that

$$\lim_{n\to\infty} \mathbf{P}\{|\tau_n| < t\} = \Phi(t),$$

so that for sufficiently large n the probability of the event $(|\tau_n| < t)$ is approximately equal to $\Phi(t)$. Consequently the probability of the event $(|\bar{\xi}_n - a| < \varepsilon)$ in which we are interested is also approximately equal to $\Phi(t)$, where $t = \varepsilon\sqrt{n}/\sigma$. Having chosen a probability P which is sufficiently near to 1, we look in the tables of the

† If the integral is over a finite interval then it is obvious that α_n tends to zero as $n \to \infty$: for improper integrals we require special estimates which we take for granted.

normal integral to find the value of $t = t(P)$ satisfying the relation $\Phi(t) = P$, and thus obtain a reliability interval for the mean value $\bar{\xi}_n$:

$$|\bar{\xi}_n - a| < \varepsilon = t(P)\frac{\sigma}{\sqrt{n}} \quad \text{with reliability } P. \tag{5.18}$$

The deviation of experimental mean values from the mathematical expectation

Let us suppose that we are interested in a *single* random variable ξ with centre a and variance σ^2, and that a sufficiently large number, n, of independent trials is carried out resulting in a set of particular values of ξ. From the experimental values $x_1, x_2, \ldots, x_n$ (however they may have been obtained) we evaluate the arithmetic mean

$$\bar{x} = \frac{x_1 + x_2 + \cdots + x_n}{n},$$

and we can assert that the inequality

$$|\bar{x} - a| < t(P)\frac{\sigma}{\sqrt{n}} \tag{5.19}$$

will be satisfied with reliability P, i.e. that $\bar{x}$ will lie in the reliability interval $(a - \varepsilon, a + \varepsilon)$, where $\varepsilon = t(\mathbf{E})\,\sigma/\sqrt{n}$. This assertion follows from the estimate (5.18) if we describe the outcome of the kth trial by a random variable ξ_k with the same probability distribution as ξ: thus x_k will be a particular value of ξ_k, and $\bar{x}$ will be a particular value of $\bar{\xi}_n$ (the independence of the random variables ξ_k is assured by the assumption that the trials are independent).

The deviation of sample means from population means

If we consider the possible values of an indicator attached to each element of a sample as the values of a random variable with distribution (3.2), then the sample mean, $\bar{x}$, will be a value of the arithmetic mean of these variables. It will therefore satisfy the inequality (5.19), where a is the population mean (the centre of the distribution (3.2)), and

$$\sigma = \sqrt{\left((x_1 - a)^2 \frac{M_1}{N} + (x_2 - a)^2 \frac{M_2}{N} + \cdots + (x_\nu - a)^2 \frac{M_\nu}{N}\right)}.$$

In other words we can expect that, with probability P, the sample mean $\bar{x}$ will differ from the population mean a by not more than $t(P)\,\sigma/\sqrt{n}$.

Reliability intervals for means can be used in the control and regulation of an industrial process in which it is necessary to maintain the value of some parameter (for example, the measurement of a component) within strictly defined tolerance limits. If from a sample we find that the mean value lies outside the limits of the reliability interval $(a - \varepsilon, a + \varepsilon)$, where $\varepsilon = t(P)\,\sigma/\sqrt{n}$, then it will be necessary to check whether the established process has changed. The precise working out of these, and similar, ideas has led to the creation of the special techniques known as "statistical quality control".

The idea of Lyapunov's central limit theorem

We established above that the normal distribution is the limiting distribution for normalised mean values, or (equivalently) for normalised sums of identically distributed summands.

The central limit theorem establishes a general condition under which the limiting distribution of normalised sums of mutually independent random summands will be the normal distribution. This problem, in its general form, was first stated in the works of P. L. Chebyshev but the condition which he formulated was too restrictive. A very general condition for the central limit theorem was published in 1900 by A. M. Lyapunov, after whom it is now named. He proved the sufficiency of the following two conditions:

(a) the absolute central moment of the third order of each random variable occurring in the sum is finite: i.e. $\mathbf{E}\,|\xi_k - a_k|^3 < \infty$, where $a_k = \mathbf{E}\xi_k$ $(k = 1, 2, \ldots)$;

(b) the ratio

$$\frac{\sum_{k=1}^{n} \mathbf{E}\,|\xi_k - a_k|^3}{\left[\sum_{k=1}^{n} \sigma^2(\xi_k)\right]^{3/2}} \to 0 \quad \text{as} \quad n \to \infty. \tag{5.20}$$

We remark that for identically distributed random summands the second condition is automatically satisfied since (5.20) becomes

$$\frac{n\mathbf{E}\,|\xi - a|^3}{[n\sigma^2]^{3/2}} = \frac{1}{\sqrt{n}}\,\frac{\mathbf{E}|\xi - a|^3}{\sigma^3}.$$

The sense of Lyapunov's conditions lies in the "limiting negligibility" of the individual summands, i.e. in the uniformly small influence of each individual random variable in the formation of the sum.

This is still more clearly evident in Lindeberg's more general conditions which require that the probabilities of large deviations of $|\xi_k - a_k|$ be uniformly small with respect to the variance of the sum $\sigma^2(\xi_1 + \xi_2 + \cdots + \xi_n)$. Roughly speaking this means that among the summands there must not be one whose possible deviations would dominate the deviations of the remainder.

Lyapunov's central limit theorem helps us to explain the wide applicability of the normal distribution in science and technology as the cumulative effect of a very large number of insignificantly small random causes. On the other hand, knowledge of the precise conditions under which the central limit theorem holds places strict limits on the region of applicability of the normal distribution.

In conclusion we must stress the importance of the law of large numbers and the central limit theorem to the entire theory of probability and its applications. Because of this crucial position many attempts have been made to extend them in different directions. From the time of Chebyshev down to our own day a leading role in this work has been played by Russian and Soviet mathematicians. In particular there is the work on dependent random variables which was begun by A. A. Markov (a pupil of Chebyshev) and continued by the Soviet mathematicians S. N. Bernstein, E. E. Slutskii, and others. Chebyshev's theorem, for example, remains true for a sequence of dependent random variables $\xi_1, \xi_2, \ldots$ with bounded variances, provided only that the dependence between ξ_i and ξ_k decreases sufficiently fast as $|i - k| \to \infty$.† We

† A precise condition can be given in terms of the correlation coefficient (see § 25); for the validity of Chebyshev's theorem it is sufficient that

$$r(\xi_i, \xi_k) \to 0 \quad \text{as} \quad |i - k| \to \infty$$

(see Ex. 7, p. 83).

can extend Lyapunov's central limit theorem to dependent variables by imposing similar conditions of weak dependence.

EXERCISES

(1) If ξ is a random variable taking integer values whose p.g.f. is $\pi(u)$ and whose characteristic function is $f(u)$, show that $f(u) = \pi(e^{iu})$.

(2) Evaluate the characteristic function of the negative exponential distribution whose density is

$$\varphi(x) = \begin{cases} 0 & x \leqq 0 \\ \lambda e^{-\lambda x} & x > 0. \end{cases}$$

Evaluate the centre and standard deviation.

(3) If k is a positive integer, $\varphi_k(x) = (\lambda x)^{k-1}e^{-\lambda x}\lambda/(k-1)!$ $(x > 0)$ is a density of Pearson type (2.27). Evaluate its characteristic function and show that $\varphi_k * \varphi_l = \varphi_{k+l}$ (in the sense of (2.41)).

(4) If, in the preceding example, $\lambda = kb^{-1}$, and $f_k(u)$ is the corresponding characteristic function, show that $\lim_{k\to\infty} f_k(u) = e^{ibu}$. Interpret.

(5) Prove the theorem of de Moivre–Laplace directly using the characteristic function of the frequency f_{μ_n} (5.6).

(6) Find a 0·99 reliability interval for ω_n when $p = 0{\cdot}01$, $n = 1000$. State your conclusions.

(7) A coin was tossed 12,000 times and tails appeared 6019 times. How well does this agree with the assumption that the appearance of a tail is equal to $\frac{1}{2}$ at each trial?

(8) In a large batch of machined articles, some numerical characteristic has the distribution table

Numerical value x	3·40	3·45	3·50	3·55	3·60	3·65	3·70	3·75	
Frequency M	150	380	1320	1530	970	470	100	80	5000

A random sample of 100 is selected. Estimate the sample mean with reliability $P = 0{\cdot}99$.

(9) A random sample from a batch of components had the following distribution table:

Numerical value x	3·40	3·45	3·50	3·55	3·60	3·65	3·70	3·75	
Frequency m	3	5	12	28	28	14	8	2	100

Can we assume that the mean value in this batch does not differ from the mean value of the batch in the previous example (assuming that σ is the same in both batches)?

(10) Measurements of the depth of the sea are normally distributed about an unknown expectation μ feet with standard deviation $\sigma = 20$ feet. How many independent observations need we make to determine the depth within an error of not greater than 15 feet with reliability 99%?

(11) In an established process for the manufacture of a dye, the yield of dye per batch is a random variable, normally distributed with expectation 50 lb, standard deviation 10 lb. Some modifications are made in the process and the yields of 16 batches from the modified process are given below. Assuming σ unchanged, determine, with reliability 95%, whether the modifications constitute an improvement in the process.

Batch No.	1	2	3	4	5	6	7	8	9	10	11	12	13	14	15	16
Yield	77	46	34	61	84	54	53	21	31	60	48	78	62	49	31	64

(12) Show that

$$\frac{1}{\sqrt{(2\pi)}}\int_x^\infty e^{-u^2/2}\,du \approx \frac{1}{x\sqrt{(2\pi)}}e^{-x^2/2} \tag{5.21}$$

when x is large.

(13) Re-examine Chapter IV, Exercise (5), using the de Moivre–Laplace limit theorem and (5.21) to obtain an estimate of n.

(14) We add n numbers (n large) each one rounded off to the same accuracy of 10^{-m}. We assume that the errors of the rounding off process are uniformly distributed on the interval $[-0{\cdot}5(10^{-m}), +0{\cdot}5(10^{-m})]$. Show that the absolute error does not exceed $\sqrt{(3n)}\,0{\cdot}5\,(10^{-m})$ with reliability 0·997.

(15) Prove that the Poisson distribution can be obtained from the binomial distribution as $n \to \infty$ and $p \to 0$, in such a way that $np = a$.

(16) Give an alternative derivation of the result of Exercise (15) using p.g.f.s.

(17) A piece of electronic equipment consists of 1000 components. The probability of failure of one component in the course of a year's operation is 0·001, and does not depend on the condition of the other components. What is the probability of failure of at least two components during one year?

CHAPTER VI

APPLICATIONS OF PROBABILITY TO THE THEORY OF OBSERVATIONS

§ 20. RANDOM ERRORS OF MEASUREMENT AND THEIR DISTRIBUTION

We define an error of measurement to be the difference $\tau = x - a$ between the result, x, of making a measurement and the true value, a, of the quantity being measured.

Every measurement contains some error. Even if we repeat a measurement under identical conditions we will usually obtain a different result. From these results we must derive the true value of the quantity which we are measuring. It is clear that both the raw data and any number derived from them will only give an approximate value of a. From all these approximate values we must select, in some sense, the best. Further, we must estimate the accuracy of our approximation, i.e. we must set up limits which we know (with a given probability) will not be exceeded by the deviation of the true value from the approximation.

In order to apply the methods of probability theory to the solution of these problems it is essential that the result of a measurement be a random variable with a definite distribution function. We shall establish the type of this distribution in the case of *direct measurements* (where the results of the measurements are considered without any transformation of the natural scale).

We will assume that *the results of the measuring process do not contain any systematic error*. A systematic error is brought about by some fixed disturbance in the measuring process, and the value of this error is either a constant for all measurements or changes in a known (and deterministic) way. A systematic error can therefore be removed either by adjusting the measuring apparatus or by applying a suitable correction to the resulting measurements.

After the removal of systematic errors the resulting measurements will contain irremovable, and unavoidable errors which have acquired the name of *random errors of measurements.*† These errors are made up of numerous causes which are difficult to specify, each of which contributes insignificantly to the fluctuation of the resulting measurement (for example, when we weigh an object on an analytical balance we can attribute errors to slight variations in the temperature and humidity of the air, to vibrations of the table, to particles of dust hitting the object being weighed, etc.).

Each of these causes itself gives rise to a so-called *elementary error of measurement*, and it is evident that an observed random error is a sum of such elementary errors. If we suppose that the number of elementary errors is very large, and the contribution of each to the observed error is very small,‡ then we can invoke the central limit theorem to assert that a random error of measurement is distributed more or less exactly according to the normal distribution. The analysis of numerous experiments and observations gives practical confirmation of this, for it shows that the distribution of observed random errors of measurement agrees very closely with the normal distribution, i.e. that the relative frequencies of observed errors of measurement are sufficiently close to the corresponding probabilities calculated from the normal distribution.

On the strength of the above argument we take as the fundamental postulate of the theory of errors that *for direct measurements the random error* τ *is normally distributed.*†† We shall assume that the centre of the distribution of random errors is zero (see

† Sometimes gross errors occur as a result of a breakdown in the measuring system or of incorrectly copying down numbers from the apparatus. The resulting measurements must be thrown away at once; they cannot be used in subsequent calculations.

‡ This is the so-called "hypothesis of elementary errors".

†† It is necessary to point out that errors are always expressed in a whole number of units related to the scale of measurement of the apparatus, whereas in the theory it is most convenient to consider the errors as *continuous* random variables which assumption simplifies all calculations. Furthermore, it is most convenient for the development of the theory to suppose that the errors are distributed along the entire axis, although sometimes this condition is contrary to the physical sense of the problem (for example, the weight of a body cannot be negative). In practice the unboundedness of τ does not disturb the conclusions because the probability of τ being outside some definite limits is very small.

the unbiasedness hypothesis below). Thus the probability density function of the random error τ is equal to

$$\varphi(t) = \frac{1}{\sigma\sqrt{(2\pi)}} \exp\left(-\frac{t^2}{2\sigma^2}\right), \ (-\infty < t < \infty). \quad (6.1)$$

The parameter $\sigma = \sqrt{(\mathbf{E}\tau^2)}$ is called *the mean square error* or *standard* error. It characteristises the accuracy of the measurements (or the precision of the apparatus).

Knowing the distribution of the random errors it is easy to find the distribution of the resulting measurements, since the resulting measurement ξ is simply related to the random error τ by the equation

$$\xi = a + \tau; \quad (6.2)$$

it follows from $\mathbf{E}\tau = 0$ that $\mathbf{E}\xi = a$. This is the so-called *unbiasedness hypothesis* which is related in practice with the absence of systematic errors. Thus, under the above conditions, we see that the results, ξ, of the measuring process are normally distributed with centre a and variance σ^2.

§ 21. THE SOLUTION OF TWO FUNDAMENTAL PROBLEMS IN THE THEORY OF ERRORS. ESTIMATION OF THE TRUE VALUE OF THE QUANTITY BEING MEASURED, AND ESTIMATION OF THE ACCURACY OF THE APPARATUS

Let $x_1, x_2, \ldots, x_n$ be the results of n direct, independent measurements of some constant quantity a. We assume that the possible results of all the measurements $\xi_1, \xi_2, \ldots, \xi_n$ are normally distributed with the same centre

$$\mathbf{E}\xi_k = a \quad (k = 1, 2, \ldots, n) \quad (6.3)$$

(the unbiasedness hypothesis) and the same variance

$$\mathbf{E}(\xi_k - a)^2 = \sigma^2 \quad (k = 1, 2, \ldots, n) \quad (6.4)$$

(the uniform precision of the measurements).

As we already know from Chapter IV, § 16 (p. 76), we can take the arithmetic mean of the measurements as an approximation to the constant a:

$$a \approx \bar{x} = \frac{x_1 + x_2 + \cdots + x_n}{n}. \tag{6.5}$$

Our first problem is to estimate the accuracy of the approximation (6.5).

Since the random variables ξ_k are independent and with the same normal distribution (centre a, variance σ^2), we know (see pp. 87 and 62) that the mean value

$$\bar{\xi} = \frac{\xi_1 + \xi_2 + \cdots + \xi_n}{n}$$

is also normally distributed with centre a and variance σ^2/n. The probability that the absolute value of the mean error $|\bar{\xi} - a|$ will not exceed some positive number ε is equal to

$$\mathbf{P}\{|\bar{\xi} - a| < \varepsilon\} = \Phi(t) \tag{6.6}$$

where

$$t = \frac{\varepsilon}{\sigma(\bar{\xi})} = \frac{\varepsilon\sqrt{n}}{\sigma},$$

and $\Phi(t)$ is the normal probability integral (2.23).

A probability P (near 1) is usually chosen in advance (e.g., $P = 0{\cdot}999$). Then, using the tables of the normal probability integral, we obtain the value of t satisfying the equation $\Phi(t) = P$ (e.g., when $P = 0{\cdot}999$ we find $t = 3{\cdot}291$). This gives the estimate $|\bar{\xi} - a| < \varepsilon$, where $\varepsilon = t\sigma/\sqrt{n}$. If we replace the random variables $\xi_k, \bar{\xi}$ by their experimental values $x_k, \bar{x}$, we obtain a *confidence interval estimate*† for a:

$$|\bar{x} - a| < t\frac{\sigma}{\sqrt{n}}$$

or

$$\bar{x} - t\frac{\sigma}{\sqrt{n}} < a < \bar{x} + t\frac{\sigma}{\sqrt{n}}. \tag{6.7}$$

† See footnote p. 91.

We interpret the estimate (6.7) in the following way: the prescribed number P is the probability that the interval $(\bar{\xi} - \varepsilon, \bar{\xi} + \varepsilon)$ will contain the true value a of the quantity which we are measuring (where $\varepsilon = t\sigma/\sqrt{n}$ and t satisfies $\Phi(t) = P$). P is called the *confidence coefficient* of the estimate (6.7).

The confidence interval estimate (6.7) has the essential drawback that the variance σ^2 is assumed known. If we replace the variance by its approximate value

$$\sigma^2 \approx s_n^2 = \frac{\sum_{k=1}^{n} (x_k - \bar{x})^2}{n - 1}$$

(see Chapter IV, § 16), then the probability that the interval (6.7) contains a is reduced. It turns out that we can give a precise estimate of a when σ^2 is unknown if we use the distribution of the random variable

$$\zeta = \frac{\bar{\xi} - a}{\dfrac{1}{\sqrt{n}\,.\,\sqrt{(n-1)}}\sqrt{\left(\sum_{k=1}^{n} (\xi_k - \bar{\xi})^2\right)}} \qquad (n \geqq 2)$$

instead of $\bar{\xi} - a$ (which depends on σ). If all the ξ_k are independent and have the same normal distribution with centre a (and unknown variance σ^2), then the random variable ζ has the distribution known as *Student's distribution* which has density function

$$S(t; n) = B_n \left(1 + \frac{t^2}{n - 1}\right)^{-n/2}$$

where

$$B_n = \frac{\Gamma(n/2)}{\sqrt{[\pi(n - 1)]}\,\Gamma[(n - 1)/2]}$$

(Γ is the Eulerian Gamma function).†

Thus the probability of the event $(|\zeta| < t)$ is equal to

$$\mathbf{P}\{|\zeta| < t\} = \int_{-t}^{t} S(t; n)\, dt.$$

† For the deriviation of Student's distribution see B. V. Gnedenko, *loc. cit.* § 24. We regard the number of variables here n as fixed.

From tables of this integral we can find the value of $t = t(P; n)$ satisfying the equation $\int_{-t}^{t} S(t;\ n)\ dt = P$ (where P is a prescribed probability). With this value of t the event $(|\zeta| < t)$ will therefore have probability P.

Inserting now the experimental values x_k of the variables ξ_k and recalling that the experimental value of $\sum (\xi_k - \bar{\xi})^2/(n-1)$ is the sample variance s_n^2, we obtain the desired estimate

$$\left|\frac{\bar{x} - a}{(1/\sqrt{n})s_n}\right| < t = t \quad (P; n)$$

or

$$\bar{x} - t\frac{s_n}{\sqrt{n}} < a < \bar{x} + t\frac{s_n}{\sqrt{n}} \quad \text{with confidence } P. \tag{6.8}$$

We give below a table† of values of $t = t(P; n)$ for different values of n and the commonly used values of P.

n \ P	0·95	0·99	0·999	n \ P	0·95	0·99	0·999
5	2·78	4·60	8·61	20	2·093	2·861	3·883
6	2·57	4·03	6·86	25	2·064	2·797	3·745
7	2·45	3·71	5·96	30	2·045	2·756	3·659
8	2·37	3·50	5·41	35	2·032	2·729	3·600
9	2·31	3·36	5·04	40	2·023	2·708	3·558
10	2·26	3·25	4·78	45	2·016	2·692	3·527
11	2·23	3·17	4·59	50	2·009	2·679	3·502
12	2·20	3·11	4·44	60	2·001	2·662	3·464
13	2·18	3·06	4·32	70	1·996	2·649	3·439
14	2·16	3·01	4·22	80	1·991	2·640	3·418
15	2·15	2·98	4·14	90	1·987	2·633	3·403
16	2·13	2·95	4·07	100	1·984	2·627	3·392
17	2·12	2·92	4·02	120	1·980	2·617	3·374
18	2·11	2·90	3·97	∞	1·960	2·576	3·291
19	2.10	2.88	3.92				

The values given in the last line (for $n = \infty$) coincide with the corresponding values in the table of the normal probability integral:

$$\Phi(1{\cdot}960) = 0{\cdot}95, \quad \Phi(2{\cdot}576) = 0{\cdot}99, \quad \Phi(3{\cdot}291) = 0{\cdot}999.$$

† This table has been reproduced from the book *Introduction to the Theory of Probability and Mathematical Statistics* by N. Arley and K. Buch.

This is explained by observing that as $n \to \infty$ Student's distribution tends to the normal distribution, as is clear from the limiting relation

$$\lim_{n\to\infty}\left(1+\frac{t^2}{n-1}\right)^{-n/2} = e^{-t^2/2}.$$

The computation of mean values

In order to apply the estimate (6.8) to a set of experimental data we must compute the mean value $\bar{x} = \sum x_k/n$ and the sample standard deviation $s_n = \sqrt{[\sum (x_k - \bar{x})^2/(n-1)]}$. The direct computation of $\bar{x}$ and s_n from these expressions is often very cumbersome, and it is sometimes possible to simplify considerably by applying a linear transformation to the data. Let us write

$$x_k = c + hu_k; \quad u_k = \frac{x - c}{h} \quad (k = 1, 2, \ldots, n) \tag{6.9}$$

where we have chosen c to be some value between the smallest and the largest of the x_k, and we have chosen the unit of measurement h so that the values of the u_k are whole numbers (this is always possible since the measurements are rounded off in a well-defined way related to the scale of the apparatus). Introducing (6.9) we derive the following computational formulae:

$$\bar{x} = c + h\bar{u}, \quad \text{where} \quad \bar{u} = \frac{\sum u_k}{n}; \tag{6.10}$$

$$\sum (x_k - \bar{x})^2 = \sum (hu_k - h\bar{u})^2 = h^2(\sum u_k^2 - 2\bar{u}\sum u_k + n\bar{u}^2)$$
$$= h^2(\sum u_k^2 - n\bar{u}^2),$$

so that

$$s_n = h\sqrt{\left(\frac{\sum u_k^2 - n\bar{u}^2}{n-1}\right)}. \tag{6.11}$$

EXAMPLE.

4·781	4·775	4·764	4·789
4·795	4·772	4·776	4·764
4·769	4·791	4·771	4·774
4·792	4·782	4·789	4·778
4·779	4·767	4·772	4·791

The accompanying table gives Millikan's first twenty measurements of the charge of an electron (in 10^{-10} absolute e.s.u.). In order to work with these data we construct a computational table, taking an initial value of $c = 4{\cdot}780$ and a unit of measurement $h = 0{\cdot}001$ (and for convenience in the computation we arrange the data in increasing order). We need two columns: a column of the reduced deviations (u) and a column of their squares (u^2).

The sum of the numbers in each column is shown in the last line:

$$\sum u_k = -29;$$

$$\sum u_k^2 = 1871.$$

Inserting these in (6.10) and (6.11) we obtain

$$\bar{u} = \frac{-29}{20} = -1{\cdot}45;$$

$$\bar{x} = 4{\cdot}780 - 0{\cdot}00145 = 4{\cdot}77855;$$

$$n\bar{u}^2 = 20\left(\frac{29}{20}\right)^2 = 42.$$

$$s_n = 0{\cdot}001\sqrt{\left(\frac{1871 - 42}{19}\right)} = 0{\cdot}00981.$$

We can suppose that the true charge of the electron, e (expressed in 10^{-10} absolute e.s.u.), is approximately given by

$$e \approx 4{\cdot}7786.$$

We will construct a confidence interval for e with confidence coefficient $P = 0{\cdot}99$. Using the table on p. 106 we find $t = 2{\cdot}861$ when $P = 0{\cdot}99$ and $n = 20$. Consequently, with confidence 0·99, we can assert that the true value of the charge lies between the limits

$$\bar{x} = t\frac{s_n}{\sqrt{n}} = 4{\cdot}77855 - 2{\cdot}861\frac{0{\cdot}00981}{\sqrt{20}} = 4{\cdot}7722$$

and

$$\bar{x} + t\frac{s_n}{\sqrt{n}} = 4{\cdot}77855 + 2{\cdot}861\frac{0{\cdot}00981}{\sqrt{20}} = 4{\cdot}7848,$$

or

$$4{\cdot}7722 < e < 4{\cdot}7848 \quad (P = 0{\cdot}99).$$

If we increase the confidence coefficient to $P = 0{\cdot}999$ then the corresponding value of t is increased to $t = 3{\cdot}883$ and consequently the interval for e is increased to

$$4{\cdot}7700 < e < 4{\cdot}7870 \quad (P = 0{\cdot}999).$$

In order to decrease the length of the interval we must either increase the number of observations or improve the precision of the individual measurements.

x	$u = \dfrac{x - 4{\cdot}780}{0{\cdot}001}$	u^2
4·764	−16	256
4·764	−16	256
4·767	−13	169
4·769	−11	121
4·771	−9	81
4·772	−8	64
4·772	−8	64
4·774	−6	36
4·775	−5	25
4·776	−4	16
4·778	−2	4
4·779	−1	1
4·781	1	1
4·782	2	4
4·789	9	81
4·789	9	81
4·791	11	121
4·791	11	121
4·792	12	144
4·795	15	225
Total	−29	1871

The estimation of precision of the measurements

The precision of the measurements is characterised by the quantity σ, the standard deviation of the distribution of the random errors.

The sample standard deviation $s_n = \sqrt{[\sum (x_k - \bar{x})^2/(n-1)]}$ is an approximate value of σ. In order to set up a confidence interval for σ we make use of the fact that the distribution of the random variable

$$\chi = \frac{1}{\sigma}\sqrt{\left[\sum_{k=1}^{n}(\xi_k - \bar{\xi})^2\right]}$$

depends only on n and not on a nor σ. If all the ξ_k ($k = 1, 2, \ldots, n$) are independent and have the same normal distribution with centre a and variance σ^2, then the distribution of the variable χ has the density function

$$R(t; n) = A_n t^{n-2} e^{-t^2/2} \quad (n \geqq 2;\ t \geqq 0)$$

where

$$A_n = \frac{1}{2^{(n-3)/2}\,\Gamma[(n-1)/2]}$$

(Γ is the Eulerian Γ-function).†

Thus the probability of the event ($t_1 < \chi < t_2$) is given by

$$\mathbf{P}\{t_1 < \chi < t_2\} = \int_{t_1}^{t_2} R(t; n)\,dt. \tag{6.12}$$

The integral (6.12) enables us to find the probability of the event ($s_n - \varepsilon < \sigma < s_n + \varepsilon$) in which we are interested. We shall write this event in the form

$$(s_n(1 - q) < \sigma < s_n(1 + q)) \tag{6.13}$$

where q denotes the relative error ε/s_n (we will assume $q < 1$).

Now, the event

$$(1 - q)\sqrt{\left[\frac{\sum(\xi_k - \bar{\xi})^2}{n-1}\right]} < \sigma < (1 + q)\sqrt{\left[\frac{\sum(\xi_k - \bar{\xi})^2}{n-1}\right]} \tag{6.14}$$

has the same probability as the event

$$\frac{\sqrt{(n-1)}}{1+q} < \frac{1}{\sigma}\sqrt{\left[\sum(\xi_k - \bar{\xi})^2\right]} < \frac{\sqrt{(n-1)}}{1-q}.$$

† For a derivation of the distribution of χ see B. V. Gnedenko, *loc. cit.*, § 24 and § 66.

This last event is in the form of $(t_1 < \chi < t_2)$, and its probability is therefore equal to the integral (6.12) with

$$t_1 = \frac{\sqrt{(n-1)}}{1+q}; \quad t_2 = \frac{\sqrt{(n-1)}}{1-q}.$$

Given a confidence coefficient P we can find the corresponding value $q = q(P; n)$ from the equation

$$\int_{\frac{\sqrt{(n-1)}}{1+q}}^{\frac{\sqrt{(n-1)}}{1-q}} R(t, n)\, dt = P,$$

so that the event (6.14) will have the given probability P. Replacing the random variables ξ_k by experimental values x_k we derive the confidence interval

$$s_n(1-q) < \sigma < s_n(1+q)$$

with confidence coefficient P.

REMARK. If $q > 1$, then the positivity of σ requires that the event (6.14) take the form

$$\left(0 < \sigma < (1+q)\sqrt{\left[\frac{\sum(\xi_k - \bar{\xi})^2}{n-1}\right]}\right),$$

which is equivalent to the event

$$(t_1 < \chi < \infty), \quad t_1 = \frac{\sqrt{(n-1)}}{1+q}.$$

The probability of this event is equal to

$$P = \int_{\frac{\sqrt{(n-1)}}{1+q}}^{\infty} R(t, n)\, dt.$$

We obtain a value $q = q(P, n)$ from this equation, and consequently, in the case $q > 1$, the confidence interval estimate of σ becomes

$$0 < \sigma < s_n(1+q).$$

We give below a table† of values of $q = q(P, n)$ for different values of n and the usual values of P.

n \ P	0·95	0·99	0·999	n \ P	0·95	0·99	0·999
5	1·37	2·67	5·64	20	0·37	0·58	0·88
6	1·09	2·01	3·88	25	0·32	0·49	0·73
7	0·92	1·62	2·98	30	0·28	0·43	0·63
8	0·80	1·38	2·42	35	0·26	0·38	0·56
9	0·71	1·20	2·06	40	0·24	0·35	0·50
10	0·65	1·08	1·80	45	0·22	0·32	0·46
11	0·59	0·98	1·60	50	0·21	0·30	0·43
12	0·55	0·90	1·45	60	0·188	0·269	0·38
13	0·52	0·83	1·33	70	0·174	0·245	0·34
14	0·48	0·78	1·23	80	0·161	0·226	0·31
15	0·46	0·73	1·15	0	0·151	0·211	0·29
16	0·44	0·70	1·07	100	0·143	0·198	0·27
17	0·42	0·66	1·01	150	0·115	0·160	0·221
18	0·40	0·63	0·96	200	0·099	0·136	1·185
19	0·39	0·60	0·92	250	0·089	0·120	0·162

EXAMPLE. Let us continue our study of Millikan's data (p. 107). We have already calculated the sample standard deviation $s_n = 0{\cdot}00981$. The precision of the measurements (in the sense of the standard deviation) is therefore given approximately by

$$\sigma \approx 0{\cdot}00981.$$

We will establish the confidence interval about this value with confidence coefficient $P = 0{\cdot}99$. From the table we find $q = 0{\cdot}58$ when $P = 0{\cdot}99$ and $n = 20$. We can therefore assert with confidence $P = 0{\cdot}99$ that the standard deviation of the errors of measurement is included between the numbers

$$s_n(1 - q) = 0{\cdot}00981\,(1 - 0{\cdot}58) = 0{\cdot}0041$$

and

$$s_n(1 + q) = 0{\cdot}00981\,(1 + 0{\cdot}58) = 0{\cdot}0145$$

† This table is reproduced from *Fundamental Problems in the Theory of Errors* by V. I. Romanovskii.

or, in other words,

$$0{\cdot}0041 < \sigma < 0{\cdot}0145 \quad (P = 0{\cdot}99).$$

If the confidence coefficient is increased to $P = 0{\cdot}999$ then the corresponding value of q is 0·88 and consequently the confidence interval for σ becomes

$$0{\cdot}0012 < \sigma < 0{\cdot}0185 \quad (P = 0{\cdot}999).$$

In order to decrease this interval it is necessary to increase the number of observations quite significantly. Thus if we want a confidence interval for σ with $q = 0{\cdot}1$ we must make 350 measurements for $P = 0{\cdot}99$, or 600 measurements for $P = 0{\cdot}999$.

Simplified estimates: "the rule of 3σ"

The estimates which we have been considering require the study of special distributions (Student's, χ, etc.). In practice simplified estimates are often applied, such as "the rule of 3σ". This rule states that the error in the approximate equation $a \approx \bar{x}$ does not exceed three times the standard deviation of the mean $\bar{x}$. If the value of σ is known then "the rule of 3σ" gives

$$|a - \bar{x}| < 3\sigma(\bar{x}) = 3\frac{\sigma}{\sqrt{n}},$$

and we have confidence $P = \Phi(3) = 0{\cdot}997$ (which follows from (6.7)). But "the rule of 3σ" is applied when σ is unknown. It is replaced by its approximate value s_n and "the rule of 3σ" takes the form

$$|a - \bar{x}| < 3\frac{s_n}{\sqrt{n}}; \qquad (6.15)$$

the confidence we have in (6.15) is significantly less than 0·997, and it decreases with decreasing n. In fact, if we compare the estimate (6.15) with (6.8) we find that for $n = 14$ the confidence coefficient for (6.15) is less than 0·99 (since $t(0{\cdot}99;\ 14) = 3{\cdot}01 > 3$), and for $n = 8$ it is equal to 0·98.

"The rule of 3σ" is also applied to the estimation of some other characteristics of the distribution, because it is always easier to find a standard deviation than to study the individual distributions. As an example we consider the application of "the rule of 3σ" to the estimation of the standard error of a measuring procedure. It can be shown that the standard deviation of the sample standard deviation is approximately equal to

$$\sigma\left(\sqrt{\left[\frac{\sum(\xi_k - \bar{\xi})^2}{n-1}\right]}\right) \approx \frac{s_n}{\sqrt{[2(n-1)]}} \tag{6.16}$$

(see Ex. 3). "The rule of 3σ" therefore states that

$$|\sigma - s_n| < 3\frac{s_n}{\sqrt{[2(n-1)]}}. \tag{6.17}$$

Comparison of this estimate with the estimate (6.13) shows that even for $n = 45$ the confidence coefficient of the estimate (6.16) is less than 0·99, since $q(0{\cdot}99; 45) = 0{\cdot}321 > 3/\sqrt{[2(45-1)]}$: for $n = 19$ it is only 0·98 and for $n = 7$ it is less than 0·95.

In order to increase the accuracy of "the rule of 3σ" it is necessary to have a larger number of measurements. In estimating σ we can achieve this by measuring different quantities with the same apparatus.

If $n_1, n_2, \ldots, n_m$ are the numbers of measurements of the first, second, ... mth quantity, and $s_1, s_2, \ldots, s_m$ the corresponding sample standard deviations, then "the rule of 3σ" becomes

$$|\sigma - S| < 3\frac{S}{\sqrt{[2(n-m)]}},$$

where

$$S = \sqrt{\left[\frac{(n_1-1)s_1^2 + (n_2-1)s_2^2 + \cdots + (n_m-1)s_m^2}{n-m}\right]}$$

and $n = n_1 + n_2 + \cdots + n_m$. When $n - m = 200$ the confidence coefficient of this estimate reaches 0·995.

EXERCISES

(1) If $x_1, \ldots, x_n$ are n numbers and $\bar{x} = (x_1 + \cdots + x_n)/n$, show that

$$\sum_{i=1}^{n}(x_i - \bar{x})^2 = \sum_{i=1}^{n} x_i^2 - n\bar{x}^2. \tag{6.19}$$

(2) If $\xi_1, \ldots, \xi_n$ are independent random variables, normally distributed with zero mean and variance σ^2, use (6.19) to show that the random variables χ and $\bar{\xi}_n$ are independent.

(3) *Continuation.* From the previous exercise show that

$$\sigma^2 \left\{ \frac{\sum_{i=1}^{n} (\xi_i - \bar{\xi}_n)^2}{n-1} \right\} = \frac{2\sigma^4}{n-1}.$$

(4) The lifetime of a light bulb is assumed to have a negative exponential distribution $1 - e^{-t/\lambda}$ for some positive, unknown λ (days). To estimate λ, one bulb is taken at random and its lifetime is observed to be 45 days. Show that we are 95% certain that $\lambda > 15$ days. Obtain a two-sided, 90% confidence interval for λ

$$\{\log_e 0{\cdot}05 = -3, \ \log_e 0{\cdot}95 = -0{\cdot}05, \text{ approx.}\}.$$

(5) Compare the confidence intervals for the expectation of a normal distribution when σ is known with the corresponding intervals when σ is unknown and estimated by s_n by computing the lengths of the confidence intervals when σ is known to be 20, $P = 0{\cdot}95$, and $n = 5, 10, 25$. Do the same when σ is unknown and $s_n = 20$.

(6) The accompanying table gives 58 measurements of the charge of an electron (in 10^{-10} absolute e.s.u.). With confidence coefficient $P = 0{\cdot}999$ estimate the true value e of the charge.

4·781	4·764	4·777	4·809	4·761	4·769
4·795	4·776	4·765	4·790	4·792	4·806
4·769	4·771	4·785	4·779	4·758	4·779
4·792	4·789	4·805	4·788	4·764	4·785
4·779	4·772	4·768	4·772	4·810	4·790
4·775	4·789	4·801	4·791	4·799	4·777
4·772	4·764	4·785	4·788	4·799	4·749
4·791	4·774	4·783	4·783	4·797	4·781
4·782	4·778	4·808	4·740	4·790	
4·767	4·791	4·771	4·775	4·747	

(7) The table gives 100 measurements of a quantity whose true value is a. With confidence coefficient $P = 0{\cdot}99$, estimate a, and also estimate the precision of the measurements.

Measurement x	3·18	3·20	3·22	3·24	3·26	3·28	
Frequency m	4	18	33	35	9	1	100

Hint: For the computation of the mean $\bar{x}$ and the sample variance s_n^2 take into account the frequency of each measurement. Construct two columns of (mu) and (mu^2), where $u = (x - x_0)/h$, and sum. Convenient values are $x_0 = 3{\cdot}22$; $h = 0{\cdot}02$.

(8) Using "the rule of 3σ" estimate the accuracy of the measuring process giving rise to the following table in which are presented 15 measurements of each of 10 different quantities (all carried out with the same instrument).

1st Series	2nd Series	3rd Series	4th Series	5th Series	6th Series	7th Series	8th Series	9th Series	10th Series
4·16	6·04	5·71	4·94	4·50	3·84	5·56	5·33	7·89	4·23
4·19	6·05	5·71	4·92	4·48	3·84	5·54	5·32	7·88	4·21
4·15	6·06	5·70	4·93	4·51	3·83	5·56	5·31	7·89	4·22
4·17	6·03	5·71	4·95	4·49	3·84	5·56	5·34	7·87	4·20
4·18	6·05	5·69	4·94	4·52	3·82	5·55	5·32	7·90	4·22
4·17	6·05	5·71	4·93	4·50	3·84	5·57	5·33	7·89	4·21
4·17	6·05	5·71	4·94	4·50	3·85	5·56	5·32	7·91	4·20
4·16	6·06	5·69	4·94	4·50	3·84	5·55	5·35	7·88	4·21
4·17	6·05	5·71	4·93	4·49	3·84	5·56	5·33	7·89	4·22
4·17	6·04	5·73	4·95	4·50	3·86	5·56	5·34	7·90	4·21
4·17	6·05	5·71	4·96	4·49	3·83	5·57	5·33	7·89	4·21
4·16	6·05	5·72	4·94	4·50	3·84	5·56	5·34	7·90	4·19
4·17	6·07	5·72	4·95	4·50	3·83	5·58	5·33	7·89	4·21
4·18	6·06	5·70	4·93	4·51	3·85	5·56	5·34	7·91	4·22
4·17	6·05	5·71	4·94	4·49	3·82	5·58	5·33	7·89	4·21

Hint: In computing each sample variance take into account the frequencies of the measurements: e.g., for the first column construct a computation table in the form:

x	m	$u = \dfrac{x - 4{\cdot}17}{0{\cdot}01}$	mu	mu^2
4·15	1	−2	−2	4
4·16	3	−1	−3	3
4·17	8	0	0	0
4·18	2	1	2	2
4·19	1	2	2	4
—	15	—	−1	13

CHAPTER VII

LINEAR CORRELATION

§ 22. ON DIFFERENT TYPES OF DEPENDENCE

The simplest form of relation between two variables is a functional dependence in which to each value of one variable there corresponds a well-defined value of the other variable. Such, for example, is the relation between the pressure and the volume of a gas in a container at constant temperature. If it is necessary to study the change in pressure due to simultaneous changes in the

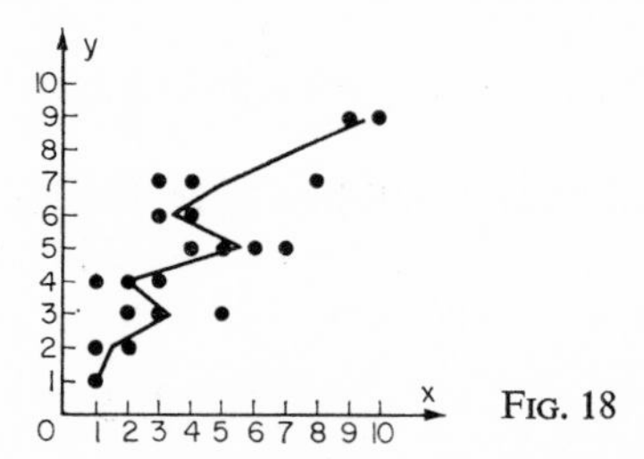

FIG. 18

volume and temperature, then we make use of the idea of a function of several variables: thus we assume that to each set of values of the independent variables there corresponds a well-defined value of a function. The study of such functional dependence constitutes a part of the calculus.

But there exist relations between physical variables which cannot be attributed to a type of functional dependence. Two examples of such relations are the connection between rainfall and the yield of a crop, or between the thickness of snow covering in winter and the volume of water in the following spring flood. In these relationships there corresponds to each value of one variable a set of possible values of the other variable. The dispersion of these possible values is explained as the influence of a

EPT. 9

very large number of additional factors which we do not take explicitly into account in studying the relationship between the given variables. Thus in practice we confine our attention mostly to studying changes in average characteristics of one variable for changes of the other. We will illustrate this by an example.

We suppose that the results of 20 experiments involving the measurement of quantities x and y are represented in Fig. 18. The change in y corresponding to a change in x can be characterised by the broken line joining the mean values of y at each value of the variable x (e.g., at $x = 1$ there are three values $y = 1, 2, 4$, whose arithmetic mean is $(1 + 2 + 4)/3 = 2\frac{1}{3}$). The dependence of the mean values on x is still a functional dependence: *to each value of x there corresponds a well-defined mean value of the variable y.*†

To study the dependence between random variables, we introduce the ideas of conditional distribution and conditional expectation.

§ 23. CONDITIONAL EXPECTATIONS AND THEIR PROPERTIES

The centre of the conditional distribution of the random variable η at the value $\xi = x$ (or the conditional expectation of η at $\xi = x$) is defined as the sum

$$\mathbf{E}_x\eta = \sum_y y\mathbf{P}\{\eta = y \mid \xi = x\}, \tag{7.1}$$

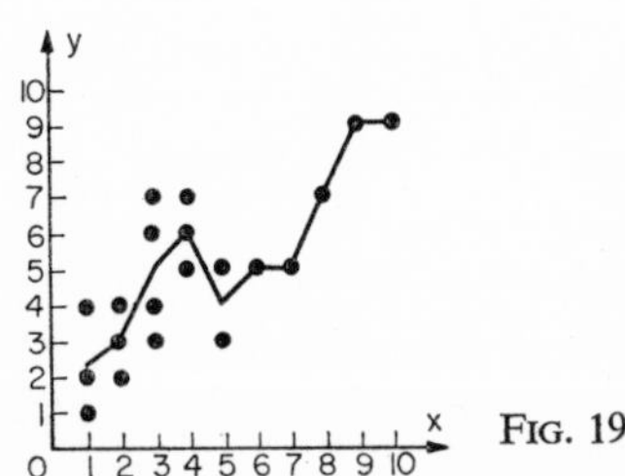

FIG. 19

† We should also record the fact that the mean dependence of x on y is another functional relation. Figure 19, using the same data as in Fig. 18, exhibits the line joining the mean values of x for each value of y (e.g., for $y = 5$ there are four values $x = 4, 5, 6, 7$, whose arithmetic mean is $(4+5+6+7)/4 = 5{\cdot}5$). Thus we can see two lines (Figs. 18 and 19), which are to be distinguished one from the other.

where $\mathbf{P}\{\eta = y|\xi = x\}$ is the conditional probability of the event $(\eta = y)$ given the event $(\xi = x)$, and the sum is taken over the range of values of η. For continuous distributions this sum is replaced by the corresponding integral

$$\mathbf{E}_x\eta = \int_{-\infty}^{\infty} y\varphi_x(y)\, dy, \tag{7.2}$$

where $\varphi_x(y)$ is the density of the conditional distribution of η given that $\xi = x$.

The conditional expectation $\mathbf{E}_x\eta$ is a function of x. This function is called *the regression function of η on ξ*: we shall denote it by $f(x)$:

$$f(x) = \mathbf{E}_x\eta.$$

The graph of the function $y = f(x)$ is called the curve of regression of η on ξ.

The derivation and study of regression functions are fundamental problems for the analysis of correlational dependence. We obtain first the important expressions

$$\mathbf{E}\eta = \mathbf{E}f(\xi), \tag{7.3}$$

$$\mathbf{E}\xi\eta = \mathbf{E}\xi f(\xi). \tag{7.4}$$

We can regard (7.4) as the generalisation of the multiplication theorem for mathematical expectations (3.9) to the case of dependent random variables. To prove (7.4) we apply the general rule for the multiplication of probabilities (1.13) to obtain

$$\mathbf{E}\xi\eta = \sum_{x,y} (xy)\, \mathbf{P}\{\xi = x, \eta = y\} = \sum_y \sum_x xy\, \mathbf{P}\{\xi = x\}\, \mathbf{P}\{\eta = y|\, \xi = x\},$$

where the sums are taken over the ranges of ξ and η. Since the term $x\,\mathbf{P}\{\xi = x\}$ does not depend on y we can rearrange this sum in the form

$$\mathbf{E}\xi\eta = \sum_x x\, \mathbf{P}\{\xi = x\} \sum_y y\, \mathbf{P}\{\eta = y|\, \xi = x\},$$

or

$$\mathbf{E}\xi\eta = \sum_x x\,\mathbf{P}\{\xi = x\}\,\mathbf{E}_x\eta = \sum_x xf(x)\,\mathbf{P}\{\xi = x\} \tag{7.5}$$

which is equal to $\mathbf{E}\xi f(\xi)$ by application of (3.11).†

(7.3) and (7.4) are special cases of the more general relation

$$\mathbf{E}u(\xi)\eta = \mathbf{E}u(\xi)f(\xi), \tag{7.6}$$

where $u(\xi)$ is any function for which the expectation $\mathbf{E}u(\xi)\eta$ exists.

We shall give a proof of (7.6) in the case when we are given the probability density function $p(x, y)$ of the two-dimensional variable (ξ, η). The probability differential $p(x, y)\,dx\,dy$ is the probability of the event $((x < \xi < x + dx)$ and $(y < \eta < y + dy))$. Applying (1.13) we obtain a relation between probability differentials

$$p(x, y)\,dx\,dy = \psi_1(x)\,dx\varphi_x(y)\,dy, \tag{7.7}$$

where $\psi_1(x) = \int_{-\infty}^{\infty} p(x, y)\,dy$ is the density function of the variable ξ and $\varphi_x(y)\,dy$ conditional probability differential of η taking a value in the interval $(y, y + dy)$ given that $\xi = x$. From (7.7) it is clear that we can write

$$\varphi_x(y) = \frac{p(x, y)}{\psi_1(x)}.$$

Substituting this expression in (7.2) we derive

$$f(x) = \mathbf{E}_x\eta = \int_{-\infty}^{\infty} y\,\frac{p(x, y)}{\psi_1(x)}\,dy.$$

A simple calculation now leads to (7.6):

$$\begin{aligned}
\mathbf{E}u(\xi)f(\xi) &= \int_{-\infty}^{\infty} [u(x)f(x)]\,\psi_1(x)\,dx \\
&= \int_{-\infty}^{\infty} u(x)\,\psi_1(x)\left[\int_{-\infty}^{\infty} y\,\frac{p(x, y)}{\psi_1(x)}\,dy\right]dx \\
&= \iint [u(x)y]\,p(x, y)\,dx\,dy = \mathbf{E}u(\xi)\eta.
\end{aligned}$$

† We remark that if the random variables ξ and η are independent then $\mathbf{P}\{\eta = y \mid \xi = x\} = \mathbf{P}\{\eta = y\}$ for all x and so $\mathbf{E}_x\eta = \mathbf{E}\eta$. Consequently (7.4) reduces to (3.9) when the variables are independent. This will also happen in the more dependent case when $f(x)$ is a constant. In fact, if $f(x) = b$ it follows from (7.3) that $\mathbf{E}\eta = \mathbf{E}b = b$, and from (7.5) that

$$\mathbf{E}\xi\eta = \sum_x xb\mathbf{P}\{\xi = x\} = b\mathbf{E}\xi = \mathbf{E}\xi\mathbf{E}\eta.$$

The relation (7.6) shows that *the mean value of any function* $u(\xi)\eta + v(\xi)$ *which is linear with respect to* η *is not changed if we replace* η *by* $f(\xi)$.

From this we derive an important minimal property of the regression function:

$$\sqrt{\{\mathbf{E}[\eta - f(\xi)]^2\}} \leqq \sqrt{\{\mathbf{E}[\eta - h(\xi)]^2\}}, \tag{7.8}$$

where $h(\xi)$ is any other function (compare (3.24)).

To prove this we write $h(\xi) - f(\xi) = u(\xi)$ and use the linearity of the mathematical expectation to obtain

$$\mathbf{E}[\eta - h(\xi)]^2 = \mathbf{E}(\eta - f(\xi) - u(\xi)]^2$$

$$= \mathbf{E}[\eta - f(\xi)]^2 + \mathbf{E}[u(\xi)]^2 - 2\mathbf{E}[\eta - f(\xi)]\, u(\xi).$$

We conclude from (7.6) that the last term is equal to zero:

$$\mathbf{E}[\eta - f(\xi)]\, u(\xi) = \mathbf{E}\eta u(\xi) - \mathbf{E}f(\xi)\, u(\xi) = 0,$$

so that

$$\mathbf{E}[\eta - h(\xi)]^2 = \mathbf{E}[\eta - f(\xi)]^2 + \mathbf{E}[h(\xi) - f(\xi)]^2 \tag{7.9}$$

from which (7.8) follows immediately.

We stress that (7.6) is concerned only with linear functions of η: replacing η by $f(\xi)$ in non-linear functions need not leave the mean value unchanged. If we look at the variance we find

$$\sigma^2(\eta) = \mathbf{E}(\eta - \mathbf{E}\eta)^2 \geqq \sigma^2[f(\xi)]. \tag{7.10}$$

This inequality follows directly from (7.9) if we take

$$h(\xi) = b = \mathbf{E}\eta = \mathbf{E}f(\xi).$$

Then

$$\mathbf{E}(\eta - b)^2 = \mathbf{E}[\eta - f(\xi)]^2 + \mathbf{E}[f(\xi) - b]^2 \geqq \sigma^2[f(\xi)].$$

We defined above the regression function of η on ξ. In the same way we can define *the regression function of* ξ *on* η:

$$\mathbf{E}_y\xi = \sum_x x\,\mathbf{P}\{\xi = x \mid \eta = y\} = g(y).$$

It follows that if the relation between ξ and η is not a strictly functional dependence then the functions $f(x)$ and $g(y)$ will not be inverse functions: i.e. the regression curve of η on ξ will not coincide with that of ξ on η.

§ 24. LINEAR CORRELATION

DEFINITION. *Two random variables* ξ *and* η *are said to be linearly correlated if both the regression functions* $f(x)$ *and* $g(y)$ *are linear.* In this case both the regression curves are straight lines: they are called the *regression lines.*

We shall derive the equation of the regression line of η on ξ, i.e. we shall determine the coefficients in the linear expression

$$f(x) = Ax + B.$$

We introduce some notation: $\mathbf{E}\xi = a$, $\mathbf{E}\eta = b$; $\mathbf{E}(\xi - a)^2 = \sigma_1^2$; $\mathbf{E}(\eta - b)^2 = \sigma_2^2$. Using (7.3) we obtain first

$$\mathbf{E}\eta = \mathbf{E}f(\xi) = \mathbf{E}(A\xi + B),$$

so that

$$b = Aa + B,$$

or

$$B = b - Aa.$$

Then applying (7.4) we find

$$\mathbf{E}\xi\eta = \mathbf{E}\xi f(\xi) = \mathbf{E}(A\xi^2 + B\xi) = A\mathbf{E}\xi^2 + (b - Aa)a,$$

whence

$$A = \frac{\mathbf{E}\xi\eta - ab}{\mathbf{E}\xi^2 - a^2} = \frac{\mathbf{E}\xi\eta - ab}{\sigma_1^2}.$$

This quantity is called *the regression coefficient of* η *on* ξ, and we shall denote it by $\varrho(\eta/\xi)$:

$$\varrho(\eta/\xi) = \frac{\mathbf{E}\xi\eta - ab}{\sigma_1^2}.$$

The equation of the regression line of η on ξ is therefore

$$y = \varrho(\eta/\xi)\,(x - a) + b. \tag{7.11}$$

In the same way we can derive the regression line of ξ on η in the form

$$x = \varrho(\xi/\eta)\,(y - b) + a, \tag{7.12}$$

where

$$\varrho(\xi/\eta) = \frac{\mathbf{E}\xi\eta - ab}{\sigma_2^2}$$

is the regression coefficient of ξ on η.

We can write the equation of the regression lines in a more symmetrical way if we introduce the dimensionless constant

$$r = \frac{\mathbf{E}\xi\eta - ab}{\sigma_1\sigma_2}. \tag{7.13}$$

This quantity is called *the coefficient of correlation between the variables ξ and η*. Note that it is symmetric in ξ and η. In this way we obtain

$$\varrho(\eta/\xi) = r\frac{\sigma_2}{\sigma_1}; \quad \varrho(\xi/\eta) = r\frac{\sigma_1}{\sigma_2},$$

and the equations of the regression lines become

$$\frac{y - b}{\sigma_2} = r\frac{x - a}{\sigma_1}, \tag{7.11*}$$

$$\frac{x - a}{\sigma_1} = r\frac{y - b}{\sigma_2}. \tag{7.12*}$$

It is clear from the equations of the regression lines that they both pass through the point (a, b), the centre of the joint distribution of the variables ξ and η. The slopes of the regression lines are given by

$$\tan\alpha = r\frac{\sigma_2}{\sigma_1}; \quad \tan\beta = \frac{1}{r}\frac{\sigma_2}{\sigma_1},$$

where the angles α and β are shown in Fig. 20.

In the following section we shall show that $|r| \leqq 1$, so that $|\tan\alpha| \leqq |\tan\beta|$. This means that the regression line of η on ξ makes a smaller angle with the x-axis than does the regression line of ξ on η. And the nearer $|r|$ is to 1, the smaller will be the angle between the regression lines. The lines will coincide if and only if $|r| = 1$. When $r = 0$ the regression lines have the equations $y = b$; $x = a$. In this case $\mathbf{E}_x\eta = b = \mathbf{E}\eta$; $\mathbf{E}_y\xi = a = \mathbf{E}\xi$.

The regression coefficients have the same sign as r, and are related by the expression

$$\varrho(\eta/\xi)\,\varrho(\xi/\eta) = r^2. \tag{7.14}$$

Since the signs of $\varrho(\eta/\xi)$ and $\varrho(\xi/\eta)$ are identical it follows that if the variable η increases (on the average) as the variable ξ increases, then the variable ξ will (on the average) increase with increasing η. The rates of increase are, however, related through the correlation coefficient.

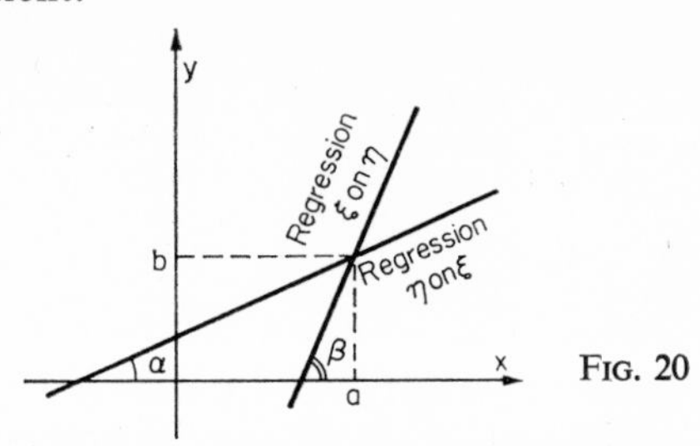

FIG. 20

EXAMPLE. *Normal correlation.* The correlation between ξ and η is called normal if the two-dimensional random variable (ξ, η) has the density function

$$p(x, y) = \frac{\sqrt{(AC - B^2)}}{2\pi} \exp\{-\tfrac{1}{2}[A(x-a)^2 + 2B(x-a)(y-b) + C(y-b)^2]\},$$

where A, B, C are constants satisfying $A > 0$, $C > 0$, $AC - B^2 > 0$.

Using (2.36) we find the marginal density function of the variable ξ to be equal to

$$\psi_1(x) = \int_{-\infty}^{\infty} p(x, y)\, dy$$

$$= \frac{\sqrt{(AC - B^2)}}{2\pi} \int_{-\infty}^{\infty} \times$$

$$\times \exp\left\{-\frac{1}{2} C \left[(y-b) + \frac{B}{C}(x-a)\right]^2 - \frac{1}{2}\left(A - \frac{B^2}{C}\right)(x-a)^2\right\} dy$$

$$= \frac{\sqrt{(AC - B^2)}}{\sqrt{(2\pi)}\sqrt{C}} \exp\left\{-\frac{1}{2}\,\frac{AC - B^2}{C}(x-a)^2\right\},$$

since

$$\frac{\sqrt{C}}{\sqrt{(2\pi)}} \int_{-\infty}^{\infty} \exp\{-\tfrac{1}{2} C(y-\lambda)^2\}\, dy = 1 \quad \text{for any } \lambda.$$

It follows that the marginal distribution of ξ is normal with centre a and variance

$$\sigma_1^2 = \frac{C}{AC - B^2}.$$

In the same way we show that the marginal distribution of η is normal with centre b and variance

$$\sigma_2^2 = \frac{A}{AC - B^2}.$$

The conditional distribution of η for a fixed value of $\xi = x$ is also normal with density function

$$\varphi_x(y) = \frac{p(x, y)}{\psi_1(x)} = \frac{\sqrt{C}}{\sqrt{(2\pi)}} \exp\left\{-\frac{1}{2} C \left[(y - b) + \frac{B}{C}(x - a)\right]^2\right\},$$

so that its centre is

$$\mathbf{E}_x \eta = b - \frac{B}{C}(x - a).$$

Analogously we show that

$$\mathbf{E}_y \xi = a - \frac{B}{A}(y - b).$$

It is clear from this that normal correlation is linear. The regression lines have the equations

$$y = -\frac{B}{C}(x - a) + b,$$

$$x = -\frac{B}{A}(y - b) + a,$$

so that the regression coefficients are

$$\varrho(\eta/\xi) = -\frac{B}{C}; \quad \varrho(\xi/\eta) = -\frac{B}{A}.$$

If we insert these expressions into (7.14) we easily obtain the coefficient of correlation

$$r = -\frac{B}{\sqrt{(AC)}}. \tag{7.15}$$

For more details about normal correlation and its significance see A. F. Mood and F. A. Graybill, *Introduction to the Theory of Statistics* (McGraw-Hill, 1963).

§ 25. THE COEFFICIENT OF CORRELATION

We shall study in greater detail the coefficient of correlation between the random variables ξ and η. It was defined in the preceding section by the expression

$$r = r(\xi, \eta) = \frac{\mathbf{E}\xi\eta - ab}{\sigma_1\sigma_2} = \frac{\mathbf{E}\xi\eta - \mathbf{E}\xi\mathbf{E}\eta}{\sigma(\xi)\sigma(\eta)}. \quad (7.13^*)$$

This coefficient measures the relative difference between the expectation of the product ($\mathbf{E}\xi\eta$) and the product of the expectations ($\mathbf{E}\xi\mathbf{E}\eta$) of the variables ξ and η. Since this difference is not zero only if the variables are dependent, we can say that the coefficient of correlation measures the degree of dependence between ξ and η.

With the help of the coefficient of correlation we can generalise the addition theorem for variances to the case of dependent variables:

$$\sigma^2(\xi + \eta) = \sigma^2(\xi) + \sigma^2(\eta) + 2r(\xi, \eta)\,\sigma(\xi)\,\sigma(\eta). \quad (7.16)$$

To prove this we recall the following expression (see p. 62):

$$\sigma^2(\xi + \eta) = \mathbf{E}(\xi - a)^2 + 2\mathbf{E}(\xi - a)(\eta - b) + \mathbf{E}(\eta - b)^2.$$

Now

$$\mathbf{E}(\xi - a)(\eta - b) = \mathbf{E}\xi\eta - a\mathbf{E}\eta - b\mathbf{E}\xi + ab = \mathbf{E}\xi\eta - ab = r\sigma_1\sigma_2,$$

and (7.16) follows directly.

Since the coefficient of correlation is dimensionless we can represent it as the expectation of the product of the normalised deviations

$$\xi_0 = \frac{\xi - a}{\sigma_1} \quad \text{and} \quad \eta_0 = \frac{\eta - b}{\sigma_2}.$$

In fact

$$\mathbf{E}\xi_0\eta_0 = \mathbf{E}\left(\frac{\xi - a}{\sigma_1} \cdot \frac{\eta - b}{\sigma_2}\right) = \frac{\mathbf{E}(\xi - a)(\eta - b)}{\sigma_1\sigma_2} = r. \quad (7.17)$$

Properties of the coefficient of correlation

THEOREM 1. *Linear transformations on the random variables ξ and η do not alter the coefficient of correlation between them:*

$$r(c_1\xi + c_2; c_3\eta + c_4) = r(\xi; \eta)$$

where c_1, c_2, c_3 and c_4 are arbitrary constants, with $c_1 > 0$, $c_3 > 0$.

This follows from the fact that the given linear transformations do not change the normalised deviations ξ_0 and η_0. Thus if $\xi' = c_1\xi + c_2 \quad (c_1 > 0)$ we have

$$\mathbf{E}\xi' = c_1\mathbf{E}\xi + c_2 = c_1 a + c_2; \quad \sigma(\xi') = c_1\sigma(\xi) = c_1\sigma_1,$$

and therefore

$$\xi_0' = \frac{\xi' - \mathbf{E}\xi'}{\sigma(\xi')} = \frac{(c_1\xi + c_2) - (c_1 a + c_2)}{c_1\sigma_1} = \frac{\xi - a}{\sigma_1} = \xi_0.$$

THEOREM 2. *The coefficient of correlation $r(\xi, \eta)$ lies between -1 and $+1$. It attains these extreme values only when there is a linear functional dependence between ξ and η.*

Proof. Since

$$\mathbf{E}\xi_0^2 = \mathbf{E}(\xi - a)^2/\sigma_1^2 = 1; \quad \mathbf{E}\eta_0^2 = 1$$

it follows from (7.17) that

$$\mathbf{E}(\xi_0 \pm \eta_0)^2 = \mathbf{E}\xi_0^2 \pm 2\mathbf{E}\xi_0\eta_0 + \mathbf{E}\eta_0^2 = 1 \pm 2r(\xi, \eta) + 1.$$

Consequently

$$1 \pm r(\xi, \eta) = \tfrac{1}{2}\mathbf{E}(\xi_0 \pm \eta_0)^2 \geqq 0, \tag{7.18}$$

and therefore

$$-1 \leqq r(\xi, \eta) \leqq +1.$$

Equality will be achieved in (7.18) if and only if $\mathbf{E}(\xi_0 \pm \eta_0)^2 = 0$; i.e. when $\xi_0 \pm \eta_0 = 0$. This last equation implies that

$$\frac{\xi - a}{\sigma_1} \pm \frac{\eta - b}{\sigma_2} = 0,$$

or

$$\eta = b \mp \frac{\sigma_2}{\sigma_1}(\xi - a).$$

THEOREM 3. *The coefficient of correlation between independent random variables is equal to zero.*

This follows directly from (3.9) and (7.13*).

It is important to remark that the converse statement is not true; i.e. that $r(\xi, \eta) = 0$ implies the independence of ξ and η. If $r(\xi, \eta) = 0$ we say that ξ and η are *uncorrelated.*

In the important case of normal correlation, $r(\xi, \eta) = 0$ does imply the independence of ξ and η. It is clear from (7.15) that in the normal case the coefficient of correlation is zero if and only if $B = 0$. But then the density function $p(x, y)$ becomes

$$p(x, y) = \frac{\sqrt{(AC)}}{2\pi} \exp \{-\tfrac{1}{2} [A(x-a)^2 + C(y-b)^2]\}$$

$$= \frac{\sqrt{A}}{\sqrt{(2\pi)}} \exp \{-\tfrac{1}{2} A(x-a)^2\} \frac{\sqrt{C}}{\sqrt{(2\pi)}} \exp \{-\tfrac{1}{2} C(y-b)^2\},$$

from which it follows immediately that the variables ξ and η are independent.

§ 26. THE BEST LINEAR APPROXIMATION TO THE REGRESSION FUNCTION

When considering linear correlation, it was comparatively easy to determine the coefficients in the regression function. With more complex correlational dependence it becomes considerably more difficult to determine the regression function. The question therefore arises of finding a linear approximation to the regression function which is in some sense the 'best linear approximation'. We established in § 23 that the function $f(\xi)$ gives the best approximation to the variable η in the sense that

$$\mathbf{E}[\eta - f(\xi)]^2 \leqq \mathbf{E}[\eta - h(\xi)]^2$$

for any function $h(\xi)$. It is reasonable therefore to take our criterion in this form and look for the function $Ax + B$ such that

$$\mathbf{E}[\eta - (A\xi + B)]^2$$

is a minimum. Our problem in this section is therefore to determine the coefficients A and B. It turns out that these coefficients are given by the same expressions as the coefficients which we found in § 24 for the case of linear correlation. We prove the following theorem.

The expression $\mathbf{E}[\eta - (A\xi + B)]^2$ *attains its unique least value when*

$$A = \varrho(\eta/\xi) = r\sigma_2/\sigma_1; \quad B = b - Aa,$$

i.e. when the linear function is

$$r\frac{\sigma_2}{\sigma_1}(\xi - a) + b.$$

In other words, among all straight lines, the regression line (7.11) gives the best approximation (in the above sense) to the actual regression of η on ξ.

Proof: Let us write $B - (b - Aa) = C$ so that

$$\mathbf{E}[\eta - (A\xi + b)]^2 = \mathbf{E}[(\eta - b) - A(\xi - a) - C]^2.$$

Using the linearity of the expectation, recalling that $\mathbf{E}(\xi - a) = 0$ and $\mathbf{E}(\eta - b) = 0$, we obtain

$$\mathbf{E}[\eta - (A\xi + B)]^2 = \mathbf{E}(\eta - b)^2 + A^2\mathbf{E}(\xi - a)^2$$

$$- 2A\mathbf{E}(\xi - a)(\eta - b) + C^2 = \sigma_2^2 + A^2\sigma_1^2 - 2Ar\sigma_1\sigma_2 + C^2. \tag{7.19}$$

In this sum the term σ_2^2 is constant, the term C^2 takes its least value (0) when $B = b - Aa$, and the term

$$A\sigma_1^2 - 2Ar\sigma_1\sigma_2 = \sigma_1^2\left(A - r\frac{\sigma_2}{\sigma_1}\right)^2 - r^2\sigma_2^2$$

takes its least value $(-r^2\sigma^2)$ when $A = r\sigma_2/\sigma_1$. The proof of the theorem is therefore complete.

Let us now evaluate $\mathbf{E}[\eta - (A\xi + B)]^2$ when $Ax + B$ is the best linear approximation. This follows immediately from (7.19) when we substitute $C = 0$ and $A = r\sigma_2/\sigma_1$:

$$\mathbf{E}\left\{\eta - \left[r\frac{\sigma_2}{\sigma_1}(\xi - a) + b\right]\right\}^2 = \sigma_2^2 - r^2\sigma_2^2 = \sigma_2^2(1 - r^2). \tag{7.20}$$

This expression enables us to clarify the sense in which the coefficient of correlation measures the degree of dependence between ξ and η. Let us rewrite (7.20) in the form

$$\sqrt{(1 - r^2)} = \sqrt{\left(\frac{\mathrm{E}\left\{\eta - \left[r\frac{\sigma_2}{\sigma_1}(\xi - a) + b\right]\right\}^2}{\sigma_2^2}\right)}$$

$$= \frac{\sigma\left\{\eta - \left[r\frac{\sigma_2}{\sigma_1}(\xi - a) + b\right]\right\}}{\sigma(\eta)}. \tag{7.21}$$

It is clear from this that $r(\xi, \eta)$ measures the relative discrepancy (in the sense of the standard deviation) between η and the best linear approximation $r\frac{\sigma_2}{\sigma_1}(\xi - a) + b$. We can therefore say that *the coefficient of correlation measures the extent of the linear dependence between ξ and η*. And the nearer r^2 approaches to 1 the less becomes the deviation of the expectation of η from the regression line of η on ξ.

Finally, we can apply the argument of this section to the regression of ξ on η.

§ 27. THE ANALYSIS OF LINEAR CORRELATION IN A GIVEN RANDOM SAMPLE. THE SIGNIFICANCE OF THE VALUE OF THE COEFFICIENT OF CORRELATION

To study the linear correlation between two variables ξ and η we carry out a sequence of independent trials (experiments, observations) from each of which we obtain a pair of numbers (x_i, y_i). We shall regard the pairs of values

$$(x_1, y_1), (x_2, y_2), \ldots, (x_n, y_n)$$

as a random sample from the range of all possible values of the variable (ξ, η). We use the method of moments (see Chapter IV) to estimate all the terms arising in the linear correlation between

ξ and η. We have, first of all, the following estimates:

$$\left.\begin{aligned} a = \mathbf{E}\xi \approx \bar{x} = \frac{\sum x}{n}; \quad b = \mathbf{E}\eta \approx \bar{y} = \frac{\sum y}{n}; \\ \sigma^2(\xi) \approx s_1^2 = \frac{\sum (x - \bar{x})^2}{n-1}; \quad \sigma^2(\eta) \approx s_2^2 = \frac{\sum (y - \bar{y})^2}{n-1}; \end{aligned}\right\} \tag{7.22}$$

$$\mathbf{E}(\xi - a)(\eta - b) \approx \frac{\sum (x - \bar{x})(y - \bar{y})}{n-1}. \tag{7.23}$$

From these we derive an approximate expression for the coefficient of correlation

$$\begin{aligned} r(\xi, \eta) \approx r_n &= \frac{\sum (x - \bar{x})(y - \bar{y})}{(n-1)s_1 s_2} \\ &= \frac{\sum (x - \bar{x})(y - \bar{y})}{\sqrt{[\sum (x - \bar{x})^2]}\sqrt{[\sum (y - \bar{y})^2]}}. \end{aligned} \tag{7.24}$$

The quantity r_n is usually called the *sample coefficient of correlation.*

Further, replacing all expectations in (7.11) and (7.12) by their sample values we obtain *the sample regression line of η on ξ:*

$$y - \bar{y} = r_n \frac{s_2}{s_1}(x - \bar{x}), \tag{7.25}$$

and *the sample regression line of ξ on η:*

$$x - \bar{x} = r_n \frac{s_1}{s_2}(y - \bar{y}). \tag{7.26}$$

The coefficients $r_n s_2/s_1$ and $r_n s_1/s_2$ are called the *sample regression coefficients.*

It is important to observe that the sample regression lines (7.25) and (7.26) possess a minimal property analogous to that discussed in § 26. We can prove in the same way that

$$\sum_{i=1}^{n} \left\{ y_i - \left[\bar{y} + r_n \frac{s_2}{s_1}(x_i - \bar{x}) \right] \right\}^2 \leqq \sum_{i=1}^{n} \{ y_i - (Ax_i + B) \}^2$$

for any other line $y = Ax + B$. The analogous property can be proved for (7.26). Figures 21 and 22 indicate the deviations we have been talking about.

It is evident from the expressions (7.22)–(7.26) that the calculation of the sample regression lines involves a large quantity of computation. As in § 21 these computations can be considerably

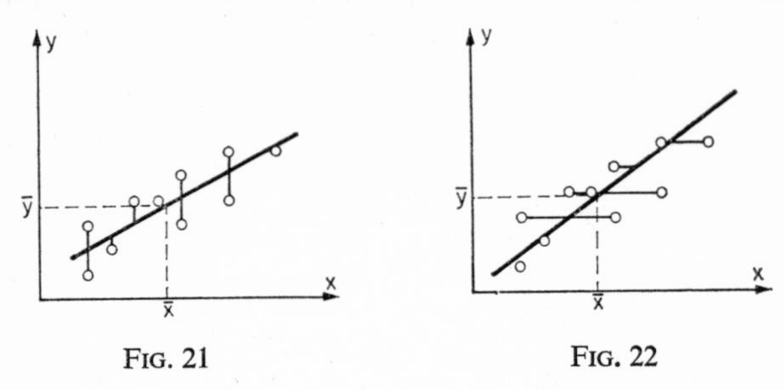

FIG. 21 FIG. 22

simplified by the use of suitable linear transformations on the quantities x and y; i.e. by the choice of convenient central values and scales of measurement. If therefore we replace x and y by the quantities

$$u = \frac{x - x_0}{h_1}; \quad v = \frac{y - y_0}{h_2} \quad (h_1 > 0, h_2 > 0),$$

we obtain the following relations (cf., § 21, p. 107):

$$x = x_0 + h_1 u; \qquad y = y_0 + h_2 v;$$

$$\bar{x} = x_0 + h_1 \bar{u} \quad \left(\bar{u} = \frac{\sum u_i}{n}\right); \quad \bar{y} = y_0 + h_2 \bar{v} \quad \left(\bar{v} = \frac{\sum v_i}{n}\right);$$

$$s_1 = h_1 \sqrt{\left[\frac{\sum u_i^2 - n(\bar{u})^2}{n-1}\right]}; \quad s_2 = h_2 \sqrt{\left[\frac{\sum v_i^2 - n(\bar{v})^2}{n-1}\right]}$$

and finally

$$r_n = \frac{\sum (h_1 u_i - h_1 \bar{u})(h_2 v_i - h_2 \bar{v})}{h_1 \sqrt{[\sum u_i^2 - n(\bar{u})^2]} h_2 \sqrt{[\sum v_i^2 - n(\bar{v})^2]}}$$

$$= \frac{\sum u_i v_i - n\bar{u}\bar{v}}{\sqrt{[\sum u_i^2 - n(\bar{u})^2]} \sqrt{[\sum v_i^2 - n(\bar{v})^2]}}.$$

N.B.

$$\sum (u_i - \bar{u})(v_i - \bar{v}) = \sum u_i v_i - \bar{u} \sum v_i - \bar{v} \sum u_i + n\bar{u}\bar{v}$$

$$= \sum u_i v_i - n\bar{u}\bar{v}.$$

EXAMPLE. *A Computation.* We shall carry out the computations on the following table (the frequency m_i indicates the number of times the pair (x_i, y_i) occurs);

x	y	Frequency m
23·0	0·48	2
24·0	0·50	4
24·5	0·49	3
24·5	0·50	2
25·0	0·51	1
25·5	0·52	1
26·0	0·49	2
26·0	0·51	1
26·0	0·53	2
26·5	0·50	1
26·5	0·52	1
27·0	0·54	2
27·0	0·52	1
28·0	0·53	3
		$n = 26$

We shall take for x the values $x_0 = 26·0$ and $h_1 = 0·5$, and for y the corresponding values $y_0 = 0·50$; $h_2 = 0·01$. Thus

$$u = \frac{x - 26·0}{0·5}; \quad v = \frac{y - 0·50}{0·01}.$$

We construct a computational table for the evaluation of the sums which we require: $\sum u$, $\sum u^2$, $\sum v$, $\sum v^2$, $\sum uv$. (Note that each term occurs in the sum with the frequency with which it occurs in the table.)

We adduce the following computational table (in the second line we have indicated an order of procedure).

x	y	m	u	um	u^2m	v	vm	v^2m	uvm
(1)	(2)	(3)	(4)	(5) = (3) × (4)	(6) = (4) × (5)	(7)	(8) = (3) × (7)	(9) = (7) × (8)	(10) = (5) × (7)
23·0	0·48	2	−6	−12	72	−2	−4	8	24
24·0	0·50	4	−4	−16	64	0	0	0	0
24·5	0·49	3	−3	−9	27	−1	−3	3	9
24·5	0·50	2	−3	−6	18	0	0	0	0
25·0	0·51	1	−2	−2	4	1	1	1	−2
25·5	0·52	1	−1	−1	1	2	2	4	−2
26·0	0·49	2	0	0	0	−1	−2	2	0
26·0	0·51	1	0	0	0	1	1	1	0
26·0	0·53	2	0	0	0	3	6	18	0
26·5	0·50	1	1	1	1	0	0	0	0
26·5	0·52	1	1	1	1	2	2	4	2
27·0	0·54	2	2	4	8	4	8	32	16
27·0	0·52	1	2	2	4	2	2	4	4
28·0	0·53	3	4	12	48	3	9	27	36
Total		26	—	−26	248	—	22	104	87

The required sums are given in the last line.†

$$\sum u = \sum{}^{*} um = -26; \quad \sum u^2 = \sum{}^{*} u^2 m = 248;$$

$$\sum v = \sum{}^{*} vm = 22; \quad \sum v^2 = \sum{}^{*} v^2 m = 104;$$

$$\sum uv = \sum{}^{*} uvm = 87.$$

From these we obtain

$$\bar{u} = \frac{-26}{26} = -1; \quad \bar{x} = 26{\cdot}0 + 0{\cdot}5(-1) = 25{\cdot}5;$$

$$s_1 = 0{\cdot}5 \sqrt{\left(\frac{248 - 26(-1)^2}{25}\right)} = 1{\cdot}49;$$

$$\bar{v} = \frac{22}{26} = 0{\cdot}846; \quad \bar{y} = 0{\cdot}50 + 0{\cdot}01 \times 0{\cdot}846 = 0{\cdot}50846;$$

$$s_2 = 0{\cdot}01 \sqrt{\left(\frac{104 - 26(0{\cdot}846)^2}{25}\right)} = 0{\cdot}0185;$$

and finally the sample coefficient of correlation

$$r_n = \frac{87 - 26(-1)\,0{\cdot}846}{\sqrt{(248 - 26)}\sqrt{(104 - 18{\cdot}6)}} = \frac{109{\cdot}0}{14{\cdot}90 \times 9{\cdot}24} = 0{\cdot}793.$$

Thus we derive the regression lines

$$y - 0{\cdot}508 = 0{\cdot}793\,\frac{0{\cdot}0185}{1{\cdot}49}(x - 25{\cdot}5) = 0{\cdot}0098(x - 25{\cdot}5),$$

$$x - 25{\cdot}5 = 0{\cdot}793\,\frac{1{\cdot}49}{0{\cdot}0185}(y - 0{\cdot}508) = 64(y - 0{\cdot}508).$$

The sample means and the regression lines are shown in Fig. 23.

Note on the reliability estimates for the coefficient of correlation

Consideration of reliability estimates for the coefficient of correlation lies outside the scope of this book. We remark only that we do not advise the use of "the rule of 3σ" since the distribution of the sample coefficient of correlation differs significantly from the normal distribution even for large n. We limit ourselves to indicating the solution to a simpler question: to what extent is it

† Where the sign Σ^* indicates that the sum is taken over the different values of the variable.

possible for the sample coefficient of correlation to differ from zero when in fact the variables ξ and η are uncorrelated?

The solution of this question is given by the probability distribution of the sample coefficient of correlation when the true value of $r(\xi, \eta)$ is zero. We quote below a table of the limits of random deviations of the quantity $|r_n| \sqrt{(n-1)}$ on condition that $r(\xi, \eta) = 0$: the limits are given as functions of the probability P and the number of observations n.† If the value of $|r_n| \sqrt{(n-1)}$ is greater than the limiting value quoted in the table, then with reliability P we can assert that the coefficient of correlation $r(\xi, \eta)$ is different from zero.

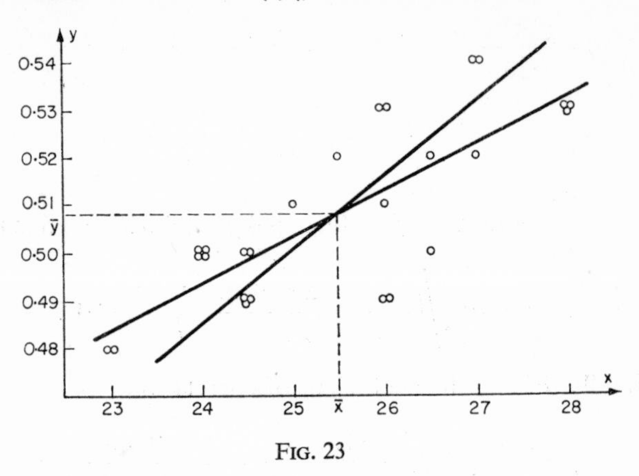

FIG. 23

TABLE OF LIMITS FOR RANDOM DEVIATIONS OF $|r_n| \sqrt{(n-1)}$.

n \ P	0·99	0·999	n \ P	0·99	0·999
10	2·29	2·62	25	2·47	3·15
11	2·32	2·68	30	2·49	3·07
12	2·35	2·73	35	2·50	3·10
13	2·37	2·77	40	2·51	3·13
14	2·39	2·81	45	2·52	3·15
15	2·40	2·85	50	2·53	3·16
16	2·41	2·87	60	2·536	3·184
17	2·42	2·90	70	2·541	3·198
18	2·43	2·92	80	2·546	3·209
19	2·44	2·94	90	2·550	3·219
20	2·45	2·96	100	2·553	3·226
			∞	2·576	3·291

† And on the additional condition that the distributions under consideration differ little from the normal. The table is reproduced from *Introduction to the Theory of Probability and Mathematical Statistics* by N. Arley and K. Buch.

EXAMPLE. In the above example

$$n = 26; \quad r_n = 0{\cdot}793; \quad |r_n| \sqrt{(n-1)} = 3{\cdot}96.$$

This number (3·96) is significantly greater than the limit for random deviations, which is approximately 3·03 with reliability $P = 0{\cdot}999$. We can therefore assert that there is correlation between the variables.

EXERCISES

(1) Find the coefficient of correlation between the variables λ_1 and λ_2 of Exercise 1, Chapter III (p. 67).

(2) The two-dimensional random variable (ξ, η) has density function $\frac{1}{2}\{f_1(x, y) + f_2(x, y)\}$ where

$$f_1(x, y) = \{2\pi(1-\varrho^2)^{1/2}\}^{-1} \exp\{-(x^2 - 2\varrho xy + y^2)/2(1-\varrho^2)\},$$
$$f_2(x, y) = \{2\pi(1-\varrho^2)^{1/2}\}^{-1} \exp\{-(x^2 + 2\varrho xy + y^2)/2(1-\varrho^2)\}.$$

Show that $r(\xi, \eta) = 0$, but that ξ, η are not independent.

(3) Calculate the sample coefficient of correlation and the sample regression lines for the following sample of observations:

x	y	Frequencies
2·3	7·1	5
2·3	7·3	4
2·6	7·3	12
2·6	7·5	8
2·6	7·7	1
2·9	7·5	5
2·9	7·7	5
3·2	7·5	4
3·2	7·7	7
3·5	7·7	2
3·5	7·9	1
3·8	7·9	1
		55

(4) A typical situation of practical importance is the following. The yield of dye in a chemical process is known to be proportional to the proportion of catalyst (temperature, pressure, etc., being kept constant). When the concentration of catalyst is x, the yield y is assumed to be of the form $y = ax + b + \varepsilon$ where a, b are unknown constants and ε is a random variable, normally distributed with zero expectation and variance σ^2. To obtain estimates of a and b the yield is measured for n different concentrations of catalyst. It is assumed that the yields $(y_1, \ldots, y_n)$ are independent random variables with the same variance, σ^2,

and it is assumed that the x_i are measured without error. Show that the estimates $\hat{a}$ and $\hat{b}$ obtained by minimising $\Sigma(y_i - ax_i - b)^2$ with respect to a and b are

$$\hat{a} = \frac{\Sigma x_i y_i - n\overline{xy}}{\Sigma (x_i - \bar{x})^2,}, \qquad \hat{b} = \bar{y} - \hat{a}\bar{x}.$$

Demonstrate that $\hat{a}$ is normally distributed with

$$\mathbf{E}\hat{a} = a \quad \text{and} \quad \sigma^2(\hat{a}) = \sigma^2/\Sigma (x_i - \bar{x})^2.$$

To obtain an estimate of σ^2, show that

$$s^2 = \Sigma (y_i - \hat{a}x_i - \hat{b})^2 = \Sigma (y_i - \bar{y})^2 - \hat{a}^2 \Sigma (x_i - \bar{x})^2,$$

and hence that $\mathbf{E}s^2 = (n - 2)\,\sigma^2$.

ANSWERS TO EXERCISES

Chapter I

(1) $(\overline{A \text{ or } B}) = (\bar{A} \text{ and } \bar{B})$. Also, $(\bar{A} \text{ and } B)$ or $(\bar{A} \text{ and } \bar{B}) = \bar{A}$. Finally, the events are clearly mutually exclusive.

(2) $B = A$ or $(B$ and $\bar{A})$, and these are mutually exclusive. So $\mathbf{P}\{B\} = \mathbf{P}\{A\} + \mathbf{P}\{B \text{ and } \bar{A}\}$.

(3) $p = \dfrac{1201}{5525} = 0{\cdot}217$.

(4) $p = \dfrac{19}{138} = 0{\cdot}138$.

(5) $p = 0{\cdot}994$.

(6) Probability that A wins $= 0{\cdot}3 + (0{\cdot}7)\,(0{\cdot}5)\,(0{\cdot}4) = 0{\cdot}44$.

(7) $1 - \left(\dfrac{5}{6}\right)^4 = 0{\cdot}5178$ (approx.): $1 - \left(\dfrac{35}{36}\right)^{24} = 0{\cdot}4913$ (approx.).

(8) $\mathbf{P}\{B\} = \mathbf{P}\{A \text{ and } B\} + \mathbf{P}\{\bar{A} \text{ and } B\}$. Then $\mathbf{P}\{\bar{A} \text{ and } B\} = \mathbf{P}\{B\} - \mathbf{P}\{A\}\,P\{B\} = \mathbf{P}\{\bar{A}\}\,P\{B\}$.

(9) $\mathbf{P}\{A|B\} \geqq \mathbf{P}\{A\}$ implies $\mathbf{P}\{A \text{ and } B\} \geqq \mathbf{P}\{A\}P\{B\}$ implies $\mathbf{P}\{B|A\} \geqq \mathbf{P}\{B\}$.

(10) $\mathbf{P}\{B|A\} = \mathbf{P}\{B|\bar{A}\}$ implies $\mathbf{P}\{B \text{ and } A\}/\mathbf{P}\{A\} = \mathbf{P}\{B \text{ and } \bar{A}\}/\mathbf{P}\{\bar{A}\}$, implies $[1 - \mathbf{P}\{A\}]\mathbf{P}\{B \text{ and } A\} = \mathbf{P}\{B \text{ and } \bar{A}\}\,\mathbf{P}\{A\}$, implies $\mathbf{P}\{B \text{ and } A\} = \mathbf{P}\{A\}\,[\mathbf{P}\{B \text{ and } A\} + \mathbf{P}\{B \text{ and } \bar{A}\}] = \mathbf{P}\{A\}\,\mathbf{P}\{B\}$.

(11) $p = \dfrac{292}{300} = 0{\cdot}973$.

(12) p_α = probability that α transmits the correct information ($\alpha = A, B, C, D$). $p_A = \dfrac{1}{3}$; $p_B = \dfrac{5}{9}$; $p_C = \dfrac{13}{27}$; $p_D = \dfrac{41}{81}$. If M_1 is the event that A told the truth, and M_2 is the event that D announced the correct information, then $\mathbf{P}\{M_2|M_1\} = p_C$, $\mathbf{P}\{M_1\} = p_A$, $\mathbf{P}\{M_2\} = p_D$, and $\mathbf{P}\{M_1|M_2\} = \mathbf{P}\{M_2|M_1\} \times \times \dfrac{p_A}{p_D} = \dfrac{13}{41}$.

Chapter II

(1)

ξ	0	1	2	3	4
p	$\dfrac{646}{1771}$	$\dfrac{1615}{3542}$	$\dfrac{285}{1771}$	$\dfrac{95}{5313}$	$\dfrac{5}{10626}$

(2)

ξ	0	1	2	3	4
p	$\frac{5}{6}$	$\frac{10}{69}$	$\frac{5}{253}$	$\frac{10}{5313}$	$\frac{1}{10626}$

(3) $\binom{n-k}{m-l}\binom{k}{l}\Big/\binom{n}{m}$. The *hypergeometric distribution.*

(6) (2.43) reduces to (2.12) when $k=2$. We may proceed by induction. Suppose (2.43) true for an experiment with k outcomes $A_1, \ldots, A_k$, and suppose $A_k = (B_1 \text{ or } B_2)$, with $P\{B_1\} = q_1$, $P\{B_2\} = q_2$. The conditional probability of m_1 occurrences of B_1 and m_2 occurrences of B_2 $(m_1 + m_2 = n_k)$ given $(n_1, n_2, \ldots, n_k)$ is $\binom{n_k}{m_1} q_1^{m_1} q_2^{m_2}/p_k^{n_k}$. Application of (1.13) gives $\dfrac{n!}{n_1! \ldots n_{k-1}!\, m_1!\, m_2!} p_1^{n_1} \ldots p_{k-1}^{n_{k-1}} q_1^{m_1} q_2^{m_2}$, so that (2.43) is true for a partition of $k+1$ outcomes.

(7) Let Y be the number of connected calls during a day, then

$P\{Y=k|X=n\} = \binom{n}{k} p^k q^{n-k}$. By (1.17), $P\{Y=k\} = \sum_n \binom{n}{k} p^k q^{n-k} e^{-\lambda}\lambda^n/n!$
$= \{e^{-\lambda}(\lambda p)^k/k!\} \sum (\lambda q)^{n-k}/(n-k)! = e^{-\lambda p}(\lambda p)^k/k!$

(8) Consider the ratio

$$\frac{\mathbf{P}\{\mu_n = m+1\}}{\mathbf{P}\{\mu_n = m\}} = \frac{n-m}{m+1} \cdot \frac{p}{q}.$$

(9) $\lim\limits_{x\to\infty} F(x) = 1$ and $\lim\limits_{x\to-\infty} F(x) = 0$ implies $A + \pi B/2 = 1, A - \pi B/2 = 0$, whence $A = \frac{1}{2}$, $B = \dfrac{1}{\pi}$. $f(x) = a/\pi(x^2 + a^2)$.

(10) $f(x, y)\,dx\,dy = (2\pi\sigma^2)^{-1} \exp\{-r^2/2\sigma^2\}.r\,dr\,d\theta = [r\sigma^{-2}\exp\{-r^2/2\sigma^2\} \times \times dr]\,[(2\pi)^{-1}\,d\theta]$, which has the factorization (2.38).

(11) $\xi/\eta = \cot\theta$. θ is uniformly distributed on $(0, 2\pi)$, so $P\{t < \cot\theta < t + \delta t\} = P\{\cot^{-1}t > \theta > \cot^{-1}(t + \delta t)\} = P\{\cot^{-1}t > \theta > \cot^{-1}t - \delta t/(1 + t^2)\} = 1/\pi(1 + t^2)$.

(12) The density is

$$\varphi(x) = \begin{cases} 0 & \text{for } x < -R, x > R, \\ \dfrac{1}{2\pi} \dfrac{1}{\sqrt{(R^2 - x^2)}} & \text{for } -R < x < R. \end{cases}$$

(13) The density is

$$\varphi(v) = \frac{1}{3v^{2/3}\sigma\sqrt{(2\pi)}} \exp\left\{-\frac{(v^{1/3} - a)^2}{2\sigma^2}\right\}.$$

(14) Since ξ^2 and η^2 are both positive, the limits of integration in (2.41) become 0 and z. When $\psi_1(y) = \psi_2(y) = (2\pi y)^{-1/2}e^{-y/2}$, (2.33), we have

$$\chi(z) = (2\pi)^{-1} \int_0^z e^{-y/2}\, y^{-1/2} e^{-(z-y)/2}\, (z - y)^{-1/2} dy$$
$$= (2\pi)^{-1} e^{-z/2} \int_0^z \{y(z - y)\}^{-1/2} dy = \tfrac{1}{2} e^{-z/2}$$

(by the substitution $y = \frac{1}{2}z(1 - \sin\theta)$), which is (2.27) with $\alpha = 1, \beta = \frac{1}{2}$.

(15) The probability density function for the sum is:

$$\varphi(x) = \begin{cases} 0 & \text{for } x < -2, x > 2; \\ \frac{1}{4}(x+2) & \text{for } -2 < x < 0; \\ \frac{1}{4}(-x+2) & \text{for } 0 < x < 2. \end{cases}$$

(16) $$\mathbf{P}\{\xi + \eta = n\} = \sum_{k=0}^{n} \mathbf{P}\{\xi = k\}\,\mathbf{P}\{\eta = n-k\}$$
$$= \frac{1}{n!}e^{-(a+b)} \sum_{k=0}^{n} \frac{n!}{k!(n-k)!} a^k b^{n-k} = e^{-(a+b)}(a+b)^n/n!$$

(17) $$\chi(z) = (2\pi)^{-1} \int_{-\infty}^{\infty} e^{-y^2/2} e^{-(z-y)^2/2}\, dy$$
$$= e^{-z^2/2}(2\pi)^{-1} \int_{-\infty}^{\infty} \exp\{-\tfrac{1}{2}(2y^2 - 2zy + \tfrac{1}{2}z^2) + \tfrac{1}{4}z^2\}\, dy$$
$$= \frac{1}{2\sqrt{\pi}} e^{-z^2/4}.$$

(18) Denote the moments of arrival of A and B by ξ and η respectively. By our assumption ξ and η are independent and uniformly distributed on the interval (0, 1): consequently the random variable (ξ, η) is uniformly distributed on the unit square (Fig. 24). The problem reduces to finding the probability of the inequality $|\xi - \eta| \leq \frac{1}{3}$; i.e. to finding the probability that (ξ, η) takes a value in the strip shown in Fig. 24 between the lines $y - x = \frac{1}{3}$ and $y - x = -\frac{1}{3}$. This probability is equal to the ratio of the area of the strip to the area of the whole square; i.e.

$$p = \frac{1 - (\frac{2}{3})^2}{1} = \frac{5}{9}.$$

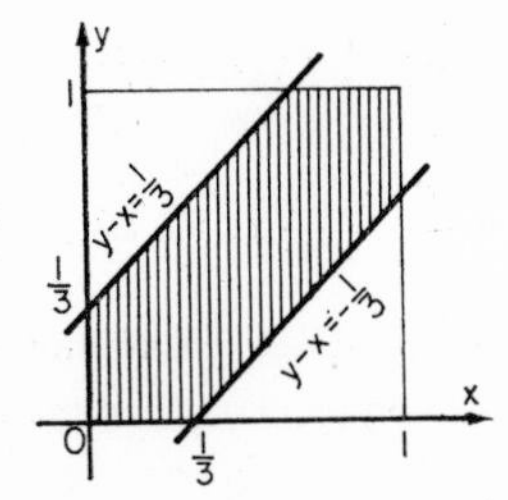

FIG. 24

(19) (a) $(pu + q)^n$; (b) $pu/(1 - qu)$; (c) $e^{-a(1-u)}$.

(20) The coefficient of u^n in $\pi_1(u)\pi_2(u)$ is $\sum_k p_k^{(1)} p_{n-k}^{(2)}$, which is just the probability that $\xi_1 + \xi_2 = n$. The random variables of Exercise (16) have p.g.f.s $e^{-a(1-u)}$, $e^{-b(1-u)}$ respectively, and their sum will have p.g.f. $e^{-(a+b)(1-u)}$.

(21) (a) $\sum \mathbf{P}\{2\xi = 2n\}\, u^{2n} = \sum \mathbf{P}\{\xi = n\}\, u^{2n} = \pi(u^2)$; (b) $\pi(u^{-1})$.

(22) If N_k is the number of shots required after the $(k-1)$th shot, up to and including the kth shot, then N_k has the p.g.f. $pu/(1 - qu)$ for each k. The

number of shots up to and including the rth hit is the random variable $N_1 + \cdots + N_r$ which has p.g.f. $\{pu/(1-qu)\}^r$. Now $\binom{-r}{n} = \binom{n+r-1}{n} \times$ $\times (-1)^n$, and the result follows by expanding $(1-qu)^{-r}$.

(23) By (19c) above, ξ_t has p.g.f. $e^{-at(1-u)}$; by (21b), $-\eta_t$ has p.g.f. $e^{-at(1-u^{-1})}$. Hence $\xi_t - \eta_t$ has p.g.f. $e^{-2at}e^{(u+u^{-1})at}$, and the coefficient of u^n is $e^{-2at} \sum_{r=0}^{\infty} \frac{(at)^{n+2r}}{r!(n+r)!}$; this infinite series is that of the modified Bessel function $I_n(2at)$. This holds also for negative n, giving the probabilities associated with excess of passengers over taxis.

(24) The result follows from the identity $\binom{2n}{n} 2^{-2n} = (-1)^n \binom{-\frac{1}{2}}{n}$.

(25) Since the random variables $\xi_1, \xi_2, \ldots$ are independent, an event depending on $\xi_1, \xi_2, \ldots, \xi_k$ will be independent of an event depending on ξ_{k+1}, $\xi_{k+2}, \ldots$. Further, if $\zeta_{2r} = 0$, the sequence $\zeta_{2r+1}, \zeta_{2r+2}, \ldots$ will have the same probabilities as $\zeta_1, \zeta_2, \ldots$. The event $(\zeta_{2n} = 0)$ may occur in the following mutually exclusive ways:

($\zeta_k = 0$ for the first time at $k = 2n$) or ($\zeta_k = 0$ for the first time at $k = 2n - 2$, and ζ_k returns to 0 at $k = 2n$) or ... or ($\zeta_k = 0$ for the first time at $k = 2$, and ζ_k returns to 0 at $k = 2n$). In view of the preceding remarks the event ($\zeta_k = 0$ for the 1st time at $k = 2r$) is independent of the event (ζ_{2r+k} returns to 0 at $k = 2n - 2r$), and the recurrence relation follows. We readily identify the right side as the coefficient of u^{2n} in $\varphi(u)\,\gamma(u)$ for $n \geqq 1$. Since $g_0 = 1$ we obtain $\gamma(u) = 1 + \varphi(u)\,\gamma(u)$. Using the result of the preceding exercise we obtain (2.44).

$f_{2n} = \frac{1}{2n-1}\binom{2n-1}{n} 2^{-2n+1}$. Note that $\varphi(1) = 1$ so that return to zero for some n is certain.

Chapter III

(1) $$\mathbf{E}(\lambda_1\lambda_2) = \frac{25}{100} \cdot \frac{24}{99}; \quad \mathbf{E}\lambda_1\mathbf{E}\lambda_2 = \left(\frac{25}{100}\right)^2.$$

(2) Make use of the independence assumption $\varphi(x, y) = \psi_1(x)\psi_2(y)$ and reduce the double integral $\iint (xy)\psi_1(x)\psi_2(y)\,dy\,dx$ to two simple integrals.

(3) $$\mathbf{E}\xi = \frac{1+2+3+4+5+6}{6} = 3{\cdot}5; \quad \sigma = \sqrt{\frac{35}{12}} = 1{\cdot}71.$$

(4) $$\mathbf{E}\xi = 7; \quad \sigma = \sqrt{\frac{70}{12}} = 2{\cdot}42.$$

Hint: Use the addition theorems for expectations and variances.

(5) $$\mathbf{E}\xi = \frac{5}{6}.$$

Hint: Represent ξ as the sum of indicator random variables.

(6)
$$\mathbf{E}\xi = \sum_{n=1}^{\infty} n(1-p)^{n-1} p = \frac{1}{p};$$

$$\sigma^2 = \mathbf{E}\xi^2 - (\mathbf{E}\xi)^2 = \sum_{n=1}^{\infty} n^2(1-p)^{n-1}p - \frac{1}{p^2} = \frac{1-p}{p^2}.$$

For $p = \frac{1}{10}$, $\mathbf{E}\xi = 10$. If the probability of a hit with each shot is $\frac{1}{10}$ then on the average we may expect to take 10 shots before the first hit.

Hint: Use the power series expansions for $1/(1-q)^2$ and $1/(1-q)^3$.

(7) $\mathbf{E}\{\xi_t - \eta_t\} = \mathbf{E}\xi_t - \mathbf{E}\eta_t = at - at = 0$. Since ξ_t, η_t are independent, $\sigma^2(\xi_t - \eta_t) = \sigma^2(\xi_t) + \sigma^2(\eta_t) = 2at$. Thus the expected excess of taxis over passengers is zero but since the standard deviation is increasing as $t^{1/2}$ we will find, with positive probability, long queues of taxis (or passengers).

(8) $\mathbf{E}\zeta = 2R/3$.

(9) $\mathbf{E}\xi = 0$, $\mathbf{E}\xi^2 = R^2/4$.

(10)
$$\mathbf{E}\xi = \int_0^{\infty} x \frac{\beta^\alpha}{\Gamma(\alpha)} x^{\alpha-1} e^{-\beta x}\, dx = \frac{\beta^\alpha}{\Gamma(\alpha)} \cdot \frac{\Gamma(\alpha+1)}{\beta^{\alpha+1}} = \frac{\alpha}{\beta};$$

$$\sigma^2 = \mathbf{E}\xi^2 - (\mathbf{E}\xi)^2 = \frac{\beta^\alpha}{\Gamma(\alpha)} \cdot \frac{\Gamma(\alpha+2)}{\beta^{\alpha+2}} - \frac{\alpha^2}{\beta^2} = \frac{\alpha}{\beta^2}.$$

Hint: Integrate by parts, or make use of the basic properties of the Γ-function.

(11)
$$\begin{aligned}\int_{-\infty}^{0} x f(x)\, dx &= \int_{-\infty}^{0} x\, dF(x) - \int_0^{\infty} x\, d[1 - F(x)] \\ &= x F(x) \big|_{-\infty}^{0} - \int_{-\infty}^{0} F(x)\, dx - x[1 - F(x)] \big|_0^{\infty} \\ &\quad + \int_0^{\infty} [1 - F(x)]\, dx.\end{aligned}$$

(12) $\mathbf{E}v = 3a\sigma^2 + a^3$.

(14)
$$\frac{\mathbf{E}(\mu_n - np)^3}{\sigma^3(\mu_n)} = \frac{npq(q-p)}{(npq)^{3/2}} = \frac{q-p}{\sqrt{(npq)}}.$$

Hint: Calculate first the third order central moment of an indicator random variable λ:

$$\mathbf{E}(\lambda - p)^3 = (1-p)^3 p + (0-p)^3 q = pq(q-p),$$

then use the addition theorem for third order central moments.

(15) To calculate the third order central moment, express it in terms of the primitive moments.

(16)
$$\mathbf{E}(\xi - a)^4 = \frac{1}{\sigma\sqrt{(2\pi)}} \int_{-\infty}^{\infty} (x-a)^4 \exp\left(-\frac{(x-a)^2}{2\sigma^2}\right) dx = 3\sigma^4.$$

Hint: Write $x - a = t\sigma$ and integrate by parts.

(17) $\sin\gamma = \sin\gamma_0 + (\gamma - \gamma_0)\cos\gamma_0 - \frac{1}{2}(\gamma - \gamma_0)^2 \sin\gamma_0 + \ldots$, so that $\mathbf{E}\{\sin\gamma\} = \sin\gamma_0 - \frac{1}{2}\mathbf{E}\{(\gamma - \gamma_0)^2\}\sin\gamma_0 + \ldots$ It follows from Example 2,

p. 63, that $\sigma^2(\gamma) = \frac{1}{3}\alpha^2$ so that $\mathbf{E}\{S\} \approx \frac{1}{2}ab(1 - \frac{1}{6}\alpha^2)\sin\gamma_0$. The exact value is $\mathbf{E}\{S\} = \frac{1}{2}ab\left(\frac{\sin\alpha}{\alpha}\right)\sin\gamma_0$.

(18) A form of Taylor's Theorem states that if $g(x)$ has 3 derivatives, then there is a point τ lying between μ and x such that $g(x) = g(\mu) + (x - \mu)g'(\mu) + \frac{1}{2}(x - \mu)^2 + \frac{1}{6}(x - \mu)^2 g'''(\tau)$. Replacing x by ξ (and observing that τ is a random variable) the first part follows immediately. Then $\sigma^2\{g(\xi)\} \approx \mathbf{E}\{g(\xi) - g(\mu)\}^2 \approx [g'(\mu)]^2\sigma^2 +$ terms involving $g''(\mu)$, etc.

(19) $\pi(u) = \Sigma\, u^n p_n$. $\pi'(u) = \Sigma\, nu^{n-1}p_n$, so $\pi'(1) = \Sigma\, np_n = \mathbf{E}\xi$. Further, $\pi''(u) = \Sigma\, n(n-1)\, u^{n-2}p_n$ and $\pi''(1) = \Sigma\, n(n-1)\, p_n = \mathbf{E}\{\xi)\xi - 1)\}$.

(a) $\pi(u) = (pu + q)^n$; $\pi'(u) = np(pu + q)^{n-1}$; $\pi''(u) = n(n-1)\,p^2 \times \times (pu + q)^{n-2}$. So $\pi'(1) = np$, $\pi''(1) = n(n-1)\,p^2$. $\sigma^2(\mu_n) = \pi''(1) + \pi'(1) - [\pi'(1)]^2 = np(1 - p)$.

(b) $\pi(u) = pu/(1 - qu) = -(p/q) + (p/q)/(1 - qu)$; $\pi'(u) = p/(1 - qu)^2$; $\pi''(u) = 2pq/(1 - qu)^3$. So $\pi'(1) = 1/p$; $\pi''(1) = 2q/p^2$. $\sigma^2 = 2q/p^2 + 1/p - (1/p)^2 = q/p^2$.

(c) $\pi(u) = e^{-a(1-u)}$; $\pi'(u) = ae^{-a(1-u)}$; $\pi''(u) = a^2e^{-a(1-u)}$. $\pi'(1) = a$; $\pi''(1) = a^2$. $\sigma^2 = a^2 + a - a^2 = a$.

(20) $\varphi(u) = 1 - (1 - u^2)^{1/2}$. $\varphi'(u) = -\frac{1}{2}(1 - u^2)^{-1/2}(-2u)$. $\varphi'(1) = \infty$.

Chapter IV

(1) The average number of incorrect junctions per minute is $\bar{x} = 2$. The sample variance is $s_n^2 = 2{\cdot}1 \approx x$. The Poisson distribution for this problem is

$$\mathbf{P}\{\xi = m\} = \frac{2^m}{m!}e^{-2}. \quad (a = 2m).$$

Comparing the observed frequencies with the theoretical frequencies we have:

No. of incorrect junctions	Frequency	Relative Frequency	Poisson probability
0	8	0·1333	0·1353
1	17	0·2833	0·2707
2	16	0·2667	0·2707
3	10	0·1667	0·1804
4	6	0·1000	0·0902
5	2	0·0333	0·0361
$\geqq 6$	1	0·0167	0·0166
	60	1·0000	1·0000

(2) The mean deviation is $\bar{x} = 0{\cdot}4$: the sample variance is $s_n^2 = 2{\cdot}57$. The corresponding normal distribution has $a = 0{\cdot}4$, $\sigma = 1{\cdot}6$, and density function

$$\varphi(x) = \frac{1}{1{\cdot}6\sqrt{(2\pi)}} \exp\left(-\frac{(x-0{\cdot}4)^2}{2\times 2{\cdot}57}\right).$$

Comparison of the observed and theoretical values gives:

Deviation x	Frequency m	Cumulative relative frequency	$F(x+0{\cdot}5)$
-3	3	0·03	0·0349
-2	10	0·13	0·1175
-1	15	0·28	0·2869
0	24	0·52	0·5249
1	25	0·77	0·7541
2	13	0·90	0·9053
3	7	0·97	0·9736
4	3	1·00	0·9948
	100		

In constructing the relative frequency we should take each value of x at the middle of a corresponding interval; for example, the frequency $m = 10$ relates to the interval $-2{\cdot}5 < x < -1{\cdot}5$. We take this into account in comparing it with the normal distribution which we take in the form

$$F(x+0{\cdot}5) = \frac{1}{2} + \frac{1}{2}\Phi\left(\frac{x+0{\cdot}5-a}{\sigma}\right),$$

where Φ is the normal probability integral.

(3) $$\bar{x} = 2{\cdot}0; \quad s_n^2 = 1{\cdot}0.$$

We recall that for a Pearson distribution $a = \alpha/\beta = 2$; $\sigma^2 = \alpha/\beta^2 = 1$, so that $\alpha = 4$; $\beta = 2$. The density of the corresponding Pearson distribution is therefore

$$\varphi(x) = \frac{16}{3!} x^3 e^{-2x}.$$

The coefficients of variation and skewness are

$$C_v = \frac{\sigma}{a} = 0{\cdot}5; \quad C_s = 1{\cdot}09 \approx 2C_v.$$

Comparing the observed with the theoretical distribution we have:

x	m	Cumulative relative frequency	$F(x+0{\cdot}5)$
0	1	0·01	0·0190
1	33	0·34	0·3527
2	41	0·75	0·7350
3	18	0·93	0·9183
4	5	0·98	0·9788
5	1	0·99	0·9951
6	1	1·00	0·9990

$$F(x+0{\cdot}5) = \frac{16}{3!}\int_0^{x+0\cdot5} x^3\, e^{-2x}\, dx.$$

(4) $\mathbf{P}\{|\xi - 3{\cdot}5| > 3\} \leqq \dfrac{35}{12\cdot 3^2} \approx \dfrac{1}{3}$.

(5) $\sigma(\omega_n) = \sqrt{(pq/n)} = \frac{1}{2}n^{-1/2}$, so $\mathbf{P}\{|\omega_n - \frac{1}{2}| > 0{\cdot}01\} \leqq 2500/n$. $n = 250{,}000$. In the next chapter we will derive a more precise estimate.

(6) $(\xi_1 + \cdots + \xi_n)^2 = \sum_1^n \xi_i^2 + 2\sum_{i>j} \xi_i\xi_j$. If $\varrho_{ij} = \mathbf{E}\{\xi_i\xi_j\}$, $i \neq j$, then $\varrho_{ij} = 0$ for $|i-j| \neq 1$ and therefore $\mathbf{E}\bar{\xi}_n^2 = n^{-2}\{\sum_1^n \mathbf{E}\xi_i^2 + 2\sum_{i=1}^{n-1} \varrho_{i,i+1}\}$. Since $|\varrho_{ij}| \leqq 1$, the second term is less, in absolute value, than $2(n-1)$. Hence $\mathbf{E}\bar{\xi}_n^2 \leqq (3n-2)/n^2 \leqq 3/n$.

(7) $\mathbf{E}\bar{\xi}_n^2 = n^{-2}\{\sum_1^n \mathbf{E}\,\xi_i^2 + 2\sum_{i>j} \varrho_{ij}\}$. Now given $\varepsilon > 0$ we can find N such that $|\varrho_{ij}| < \varepsilon$ for $|i-j| \geqq N$. For $n > N$, the absolute value of the second sum is of the form $|\sum_{j=1}^{n} \{\sum_{i=j}^{j+N} \varrho_{ij} + (n-j-N)\,\varepsilon\} \leqq nN + \frac{1}{2}(n-N)^2\varepsilon$. Hence $\mathbf{E}\bar{\xi}_n^2 \leqq \dfrac{N+1}{n} + \frac{1}{2}\left(1 - \dfrac{N}{n}\right)^2 \varepsilon$, and since ε is arbitrary the result follows.

(8) $\int x f(x)\, dx = 0$ for $\alpha > 1$. $\int x^2 f(x)\, dx = \begin{cases} \alpha/(\alpha-2) & \text{for } \alpha > 2 \\ \infty & \text{for } \alpha \leqq 2. \end{cases}$

Chapter V

(1) $\pi(u) = \sum p_n u^n; f(u) = \sum p_n e^{iun} = \sum p_n (e^{iu})^n$.

(2) $\lambda \int_0^\infty e^{-\lambda x} e^{iux}\, dx = \lambda \int_0^\infty e^{-(\lambda - iu)x}\, dx = \lambda/(\lambda - iu)$. A full account of the evaluation of this integral would require a discussion of related contour integrals in the complex plane. The plausible answer is the correct one so we by-pass its complete justification. This applies also to the next exercise.

(3) $\int_0^\infty e^{iux}\{e^{-\lambda x}(\lambda x)^{k-1}/k!\}\,\lambda\, dx = (\lambda^k/k!)\int_0^\infty e^{-(\lambda-iu)x}\, x^{k-1}\, dx$

$= [\lambda/(\lambda - iu)]^k (1/k!) \int_0^\infty e^{-(\lambda - iu)x}[(\lambda - iu)\, x]^{k-1}(\lambda - iu)\, dx = [\lambda/(\lambda - iu)]^k$.

By (5.4), $[\lambda/(\lambda - iu)]^{k+l} = [\lambda/(\lambda - iu)]^k[\lambda/(\lambda - iu)]^l$ is equivalent to $\varphi_{k+l} = \varphi_k * \varphi_l$. Note that $\varphi_k = \varphi_{k-1} * \varphi_1 = \varphi_{k-2} * \varphi_1 * \varphi_1$, etc., so that if $\varphi^{(2)} = \varphi_1 * \varphi_1$, $\varphi^{(3)} = \varphi^{(2)} * \varphi_1, \ldots, \varphi^{(k)} = \varphi^{(k-1)} * \varphi_1$, then $\varphi_k = \varphi^{(k)}$. Thus if $\tau_1, \ldots, \tau_k$ are independent random variables, identically distributed with density φ_1, $\tau = \tau_1 + \cdots + \tau_k$ will be distributed with density φ_k. This provides a model for service in a class of queuing systems. For example, if a mechanic servicing an engine takes a time τ_1 to service the electrical system, τ_2 to service the fuel system, etc., then the total time taken to service the entire engine is the sum of these.

(4) $[kb^{-1}/(kb^{-1} - iz)]^k = [1 - (izb/k)]^k \to e^{izb}$. The right side is the characteristic function of a random variable taking the value b with probability 1.

(5) Using (5.6), (5.5), and (5.11) we can obtain the characteristic function $f_n(u)$ of the normalised frequency $\tau_n = (\mu_n - np)/\sqrt{(npq)}$, and then, expanding the exponential functions

$$\exp\left(i \frac{u}{\sqrt{n}} \sqrt{\frac{q}{p}}\right) \quad \text{and} \quad \exp\left(-i \frac{u}{\sqrt{n}} \sqrt{\frac{p}{q}}\right),$$

we can write the characteristic function in the form

$$f_n(u) = \left(1 - \frac{u^2}{2n} - i \frac{u^3}{3n\sqrt{n}} \cdot \frac{q-p}{\sqrt{(pq)}} + \cdots\right)^n.$$

(6)

$$|\omega_n - p| < 2{\cdot}576 \sqrt{\frac{0{\cdot}0099}{1000}} = 0{\cdot}0081,$$

i.e. $0{\cdot}0019 < \omega_n < 0{\cdot}0181$. With reliability 0·99 we can expect that in 1000 repeated trials the event in which we are interested will occur between 2 and 18 times: $(2 < \mu_n = n\omega_n < 18)$.

(7) On this assumption the probability of observing a greater deviation is equal to

$$\mathbf{P}\left\{|\omega_n - \tfrac{1}{2}| \geqq \frac{19}{12{,}000}\right\} \approx 1 - \Phi\left(\frac{\frac{19}{12{,}000}\sqrt{12{,}000}}{\sqrt{(0{\cdot}5 \times 0{\cdot}5)}}\right) = 0{\cdot}738.$$

This probability is not small so that the result of this experiment gives us no cause to doubt the hypothesis concerning the probability of obtaining a tail.

(8) The population mean is $a = 3{\cdot}55$; $\sigma = 0{\cdot}05\sqrt{1{\cdot}844} = 0{\cdot}068$. The estimate of the sample mean for $P = 0{\cdot}99$ is

$$|\bar{x} - a| < 2{\cdot}576 \frac{0{\cdot}068}{\sqrt{100}} = 0{\cdot}0175,$$

or

$$3{\cdot}5325 < \bar{x} < 3{\cdot}5675.$$

(9) Here $\bar{x} = 3{\cdot}55 + 0{\cdot}05 \dfrac{57}{100} = 3{\cdot}55 + 0{\cdot}0285$. The deviation from the population mean of the previous example is $\varepsilon = 0{\cdot}0285$. The probability of a deviation at least as large as this is equal to

$$\mathbf{P}\{|\bar{x} - a| \geqq 0{\cdot}0285\} \approx 1 - \Phi\left(\frac{0{\cdot}0285\sqrt{100}}{0{\cdot}068}\right) < 0{\cdot}00003.$$

This probability is very small so that it is not possible to assume that the sample was taken from a batch with mean value $a = 3{\cdot}55$ (with $\sigma = 0{\cdot}068$).

(10) If $\bar{\xi}_n$ is the average of n independent observations, we seek n so that $\mathbf{P}\{|\bar{\xi}_n - \mu| > 15\} < 0{\cdot}01$. $\sigma(\bar{\xi}_n) = \sigma n^{-1/2}$, whence $\mathbf{P}\left\{\left|\frac{\bar{\xi}_n - \mu}{\sigma n^{-1/2}}\right| > \frac{15n^{1/2}}{\sigma}\right\} < 0{\cdot}01$. From Table II we find $15n^{1/2}/\sigma = 2{\cdot}58$. Since $\sigma = 20$ is given we find $n = 11{\cdot}83$. Thus we will attain the accuracy required with 12 or more measurements.

(11) If the modifications do not constitute an improvement, $\bar{x}$ is normally distributed with $a = 50$, $\sigma = 10/\sqrt{16}$. And since we are not interested in values of $\bar{x}$ less than 50 (modifications which cannot do better than that would be straightway rejected), we can set up a one-sided reliability interval for $\bar{x}$ by seeking $t_1(P)$ such that $\mathbf{P}\{\bar{x} - a < t_1(P)\,\sigma/n^{1/2}\} = P$. If $\bar{x}$ lies outside this interval (i.e. $\bar{x} > a + t_1(P)\,\sigma/n^{1/2}$) we may regard the modifications as an improvement with reliability P.† To determine $t_1(P)$, observe that it is related to $t(P)$ of (5.19): $t_1(P) = t(2P - 1)$. Hence $t_1(0{\cdot}95) = t(0{\cdot}90) = 1{\cdot}65$ (approx.), by reference to Table I. In this example, $a + t_1(P)\sigma/n^{1/2} = 50 + (1{\cdot}65)(2{\cdot}5) = 54{\cdot}125$. Since $\bar{x} = 53{\cdot}31$ is less than $54{\cdot}125$, we are 95% confident that the modifications do not constitute an improvement.

(12)
$$\begin{aligned}&(2\pi)^{-1/2}\int_x^{\infty} u^{-1} d(-e^{-u^2/2})\\ &= -(2\pi)^{-1/2}u^{-1}e^{-u^2/2}\Big|_x^{\infty} - (2\pi)^{-1/2}\int_x^{\infty} u^{-2}\,e^{-u^2/2}\,du\\ &= (2\pi)^{-1/2}x^{-1}\,e^{-x^2/2}\end{aligned}$$

plus a term which is smaller than $x^{-1}[1 - \Phi(x)]$. Successive integration by parts yields a series in decreasing powers of x which diverges for each x but whose first few terms decrease rapidly to give a useful approximation.

(13)
$$\mathbf{P}\{|\omega_n - \tfrac{1}{2}| > \varepsilon\} = \mathbf{P}\{|\tau_n| > \tfrac{1}{2}\varepsilon n^{1/2}\} \approx \frac{2}{\sqrt{(2\pi)}}\int_{1/2\varepsilon n^{1/2}}^{\infty} e^{-u^2/2}\,du$$
$$\approx \frac{2}{(\tfrac{1}{2}\varepsilon n^{1/2})\sqrt{(2\pi)}}\,e^{-(\frac{1}{2}\varepsilon n^{1/2})^2/2}.$$

Writing $x = \tfrac{1}{2}\varepsilon n^{1/2}$ we require x such that $(0{\cdot}005)x = \dfrac{1}{\sqrt{(2\pi)}}\,e^{-x^2/2}$. Reference to Table III shows that $x \approx 2{\cdot}6$. Taking $\varepsilon = 0{\cdot}01$ we obtain $n \approx 16{,}900$.

(14) The total rounding off error can be regarded as the sum of n independent, uniformly distributed random variables. We can suppose that for n sufficiently large the distribution of the error of the sum will be near to the normal distribution with centre 0 and standard deviation

$$\sigma = n^{1/2}\,\sigma(\xi) = n^{1/2}\,\frac{10^{-m}}{2\sqrt{3}}.$$

(See § 12, Example 2, p. 63).

† *Translator's note.* Among Anglo-American statisticians, this discussion is carried out under the heading of "Testing Hypotheses". $1 - P$ is then called the "Significance Level" of the test. In this case we would say that the value $\bar{x} = 53{\cdot}31$ is not significant at the 5% level.

(15) Substituting $p = a/n$ we obtain

$$\binom{n}{m} p^m q^{n-m} = \frac{n(n-1)\ldots(n-m+1)}{m!} \frac{a^m}{n^m} \left(1 - \frac{a}{n}\right)^{n-m}.$$

Now let $n \to \infty$.

(16) $(pz+q)^n = (pz+1-p)^n = [1 - a(1-z)/n]^n \to e^{-a(1-z)}$.

(17) Application of Poisson approximation to the Binomial distribution. Here $p = 0{\cdot}001$, $n = 1000$, so $a = np = 1$. The probability of failure of at least 2 components is $1 - \mathbf{P}\{\text{no failures}\} - \mathbf{P}\{1 \text{ failure}\} = 1 - 0{\cdot}368 - 0{\cdot}368 = 0{\cdot}264$ (referring to Table IV).

Chapter VI

(1) $$\Sigma (x_i - \bar{x})^2 = \Sigma x_i^2 - 2\bar{x} \Sigma x_i + n\bar{x}^2 = \Sigma x_i^2 - 2\bar{x}(n\bar{x}) + n\bar{x}^2 - n\bar{x}^2 = \Sigma x_i^2 .$$

(2) If the probability density of ξ_i is $(2\pi\sigma^2)^{-1/2} \exp\{-x_i^2/2\sigma^2\}$, the joint density of $\xi_1 \ldots \xi_n$ is $(2\pi\sigma^2)^{-n/2} \exp\{-\Sigma x_i^2/2\sigma^2\} = (2\pi\sigma^2)^{-n/2} \exp\{-\Sigma (x_i - \bar{x})^2/2\sigma^2\} \times \times \exp\{-n\bar{x}^2/2\sigma^2\}$ which is in the form (2.38). To obtain (6.12) from the first factor is beyond the scope of this book. The second factor re-establishes that $\bar{\xi}_n$ is normally distributed with expectation zero and variance σ^2/n.

(3) $\mathbf{E}\{\Sigma (\xi_i - \bar{\xi}_n)^2 + n^{-1}(\Sigma \xi_i)^2\}^2 = \mathbf{E}(\Sigma \xi_i^2)^2$. Using Exercise 16 of Chapter III, the right side is seen to be $n(n+2)\sigma^4$. Expanding the left side we have $\mathbf{E}\{\Sigma (\xi_i - \bar{\xi}_n)^2\}^2 + 2n^{-1}\mathbf{E}\{\Sigma (\xi_i - \bar{\xi}_n)^2\}\{\Sigma \xi_i\}^2 + n^{-2}\mathbf{E}\{\Sigma \xi_i\}^4$. Because of the independence the middle term becomes $\mathbf{E}\{\Sigma (\xi_i - \bar{\xi}_n)^2\} \mathbf{E}\{\Sigma \xi_i\}^2 = (n-1)\sigma^2 \times \times n\sigma^2$. Finally $\mathbf{E}\{\Sigma \xi_i\}^4 = 3n^2\sigma^4$, so that $\mathbf{E}\{\Sigma (\xi_i - \bar{\xi}_n)^2\}^2 = (n^2 - 1)\sigma^4$. The result follows with the help of (3.25). (6.16) follows by application of Exercise (17) of Chapter III.

(4) The probability of observing a lifetime greater than or equal to 45 days is $e^{-45/\lambda}$. If λ is small, the observed lifetime will be in the tail of the distribution so the probability of actually having observed a lifetime of 45 days would be small. Taking 0·05 as our prescribed small probability, we require $e^{-45/\lambda} > 0{\cdot}05$. Whence $\lambda > -45/\log_e 0{\cdot}05 = 15$ days, with reliability 0·95.

In the same way, an observed lifetime of less than 45 days would be an unreasonable observation if λ were very large. Again, taking 0·05 as our small probability, we would require $1 - e^{-45/\lambda} > 0{\cdot}05$, or $\lambda < -45/\log_e 0{\cdot}95 = 900$, with reliability 0·95. Taking these two inequalities together, we are 90% confident that $15 < \lambda < 900$ (days).

(5) σ known: 95% confidence interval $\bar{x} \pm 1{\cdot}96\, \sigma/n^{1/2}$.
σ unknown: 95% confidence interval $\bar{x} \pm t(0{\cdot}95, n)\, s_n/n^{1/2}$.

n	5	10	25
σ known	35	24·8	15·6
σ unknown	49·6	29·2	16·4

(6) $x = 4{\cdot}780 + (0{\cdot}001)\ (0{\cdot}81) = 4{\cdot}78081 \approx e;$

$$\frac{s_n}{\sqrt{n}} = 0{\cdot}001 \sqrt{\left(\frac{13{,}367 - 38}{57 \times 58}\right)} = 0{\cdot}00201\,.$$

The estimate is

$$|e - \bar{x}| < t(s_n/\sqrt{n});$$

$$t = t(0{\cdot}999;\, 58) = 3{\cdot}470; \quad \varepsilon = t(s_n/\sqrt{n}) = 0{\cdot}00697;$$

$$4{\cdot}7738 < e < 4{\cdot}7878.$$

(7) $\bar{x} = 3{\cdot}22 + (0{\cdot}02)\ (0{\cdot}3) = 3{\cdot}226 \approx a;$

$$s_n = 0{\cdot}02 \sqrt{\left(\frac{114 - 9}{99}\right)} = 0{\cdot}0206 \approx \sigma.$$

For the estimate of a:

$$t(0{\cdot}99;\, 100) = 2{\cdot}627; \quad \varepsilon = t(s_n/\sqrt{n}) = 0{\cdot}054;$$

$$|a - 3{\cdot}226| < 0{\cdot}054, \quad \text{or} \quad 3{\cdot}172 < a < 3{\cdot}280.$$

For the estimate of σ:

$$0{\cdot}0206\,(1 - q) < \sigma < 0{\cdot}0206\,(1 + q),$$

$$q(0{\cdot}99;\, 100) = 0{\cdot}198;$$

so

$$0{\cdot}0165 < \sigma < 0{\cdot}0247.$$

(8)

$$14s_1^2 = 12{\cdot}93\ \ 10^{-4}; \quad 14s_2^2 = 12{\cdot}93\ \ 10^{-4};$$

$$14s_3^2 = 15{\cdot}73\ \ 10^{-4}; \quad 14s_4^2 = 14{\cdot}93\ \ 10^{-4};$$

$$14s_5^2 = 13{\cdot}73\ \ 10^{-4}; \quad 14s_6^2 = 16{\cdot}40\ \ 10^{-4};$$

$$14s_7^2 = 15{\cdot}73\ \ 10^{-4}; \quad 14s_8^2 = 14{\cdot}93\ \ 10^{-4};$$

$$14s_9^2 = 16{\cdot}40\ \ 10^{-4}; \quad 14s_{10}^2 = 13{\cdot}73\ \ 10^{-4};$$

$$S = \sqrt{\left[\frac{14(S_1^2 + S_2^2 + \cdots + S_{10}^2)}{140}\right]} = 0{\cdot}01 \sqrt{\frac{147{\cdot}5}{140}} = 0{\cdot}01026 \approx \sigma.$$

The estimate is

$$|\sigma - S| < 3\,\frac{S}{\sqrt{2140}} = 0{\cdot}00184,$$

or

$$0{\cdot}0084 < \sigma < 0{\cdot}0121.$$

Chapter VII

(1)

$$r(\lambda_1\lambda_2) = \frac{\mathbf{E}\lambda_1\lambda_2 - \mathbf{E}\lambda_1\mathbf{E}\lambda_2}{\sigma_1\sigma_2} = \frac{\dfrac{25}{100}\times\dfrac{24}{99} - \left(\dfrac{25}{100}\right)^2}{\dfrac{25}{100}\,\dfrac{75}{100}} = -\frac{1}{99}.$$

(2) $\mathbf{E}\xi = \mathbf{E}\eta = 0$. $\mathbf{E}\xi\eta = \frac{1}{2}\iint xy\{f_1(x, y) + f_2(x, y)\}\, dx\, dy = \frac{1}{2}\{\varrho - \varrho\} = 0$. But $\frac{1}{2}\{f_1(x, y) + f_2(x, y)\}$ does not factorise into the form of (2.38).

(3)

$$\bar{x} = 2{\cdot}9 + 0{\cdot}3\left(-\frac{19}{55}\right) = 2{\cdot}796;\quad \bar{y} = 7{\cdot}5 + 0{\cdot}2\left(\frac{7}{55}\right) = 7{\cdot}475;$$

$$s_1 = 0{\cdot}3\sqrt{\frac{\left(89 - \dfrac{19^2}{55}\right)}{54}} = 0{\cdot}3\sqrt{\frac{82{\cdot}42}{54}} = 0{\cdot}371;$$

$$s_2 = 0{\cdot}2\sqrt{\frac{\left(59 - \dfrac{7^2}{55}\right)}{54}} = 0{\cdot}2\sqrt{\frac{58{\cdot}11}{54}} = 0{\cdot}208;$$

$$r_n = \frac{60 - \dfrac{19{\cdot}7}{55}}{\sqrt{(82{\cdot}42\times 58{\cdot}11)}}\,\frac{57{\cdot}58}{69{\cdot}21} = 0{\cdot}853.$$

The sample regression lines are

$$y - 7{\cdot}47 = 0{\cdot}833\,\frac{0{\cdot}208}{0{\cdot}371}(x - 2{\cdot}80) = 0{\cdot}467\,(x - 2{\cdot}80),$$

$$x - 2{\cdot}80 = 0{\cdot}833\,\frac{0{\cdot}371}{0{\cdot}208}(y - 7{\cdot}47) = 1{\cdot}49\,(y - 7{\cdot}47).$$

(4) We imitate the argument leading up to (7.19). Writing $b - (\bar{y} - a\bar{x}) = c$, we obtain

$$\Sigma\,(y_i - ax_i - b)^2 = \Sigma\,(y_i - \bar{y})^2 + c^2 + \left\{a\sqrt{\Sigma\,(x_i - \bar{x})^2} - \frac{\Sigma\,(x_i - \bar{x})\,(y_i - \bar{y})}{\sqrt{\Sigma\,(x_i - \bar{x})^2}}\right\}^2 - \{\Sigma\,(x_i - \bar{x})\,(y_i - \bar{y})\}^2/\Sigma\,(x_i - \bar{x})^2.$$

Then the first and last terms are constants and the second and third terms taken on minimum values at the values $\hat{a}$ and $\hat{b}$. $\hat{a}$ is a linear function of the y_i and so, by Example 5 of § 17, $\hat{a}$ is normally distributed. $\mathbf{E}y_i = ax_i + b$, $\mathbf{E}\bar{y} = a\bar{x} + b$, so $\mathbf{E}\hat{a} = \dfrac{\Sigma (x_i - \bar{x}) \mathbf{E}(y_i - \bar{y})}{\Sigma (x_i - \bar{x})^2} = a$. Also $\Sigma (x_i - \bar{x})(y_i - \bar{y}) = \Sigma x_i y_i - n\overline{xy}$ $= \Sigma x_i y_i - \bar{x} \Sigma y_i = \Sigma (x_i - \bar{x}) y_i$. The y_i are independent so, by (3.20′), $\sigma^2(\hat{a}) = \dfrac{\Sigma (x_i - \bar{x})^2 \sigma^2(y_i)}{\{\Sigma (x_i - \bar{x})^2\}^2} = \dfrac{\sigma^2}{\Sigma (x_i - \bar{x})^2}$. Write $y_i = ax_i + b + \varepsilon_i$. Then $\bar{y} = a\bar{x} + b + \bar{\varepsilon}$, where $\bar{\varepsilon} = (\varepsilon_1 + \cdots + \varepsilon_n)/n$. Introducing these we have $s^2 = \Sigma \{a(x_i - \bar{x}) + \varepsilon_i - \bar{\varepsilon}\}^2 - \hat{a}^2 \Sigma (x_i - \bar{x})^2 = a^2 \Sigma (x_i - \bar{x})^2 + 2a \Sigma (x_i - \bar{x}) \times$ $\times (\varepsilon_i - \bar{\varepsilon}) + \Sigma (\varepsilon_i - \bar{\varepsilon})^2 - \hat{a}^2 \Sigma (x_i - \bar{x})^2$. Now $\mathbf{E}\varepsilon_i = 0 = \mathbf{E}\bar{\varepsilon}, \mathbf{E}\, \Sigma (\varepsilon_i - \bar{\varepsilon})^2$ $= (n-1)\sigma^2$, and $\mathbf{E}\hat{a}^2 = [\sigma^2 / \Sigma (x_i - \bar{x})^2] + a^2$ and the result follows.

APPENDIX

TABLES

VALUES OF THE NORMAL PROBABILITY INTEGRAL

$$\Phi(t) = \frac{2}{\sqrt{(2\pi)}} \int_0^t e^{-x^2/2} dx; \quad \Phi(-t) = -\Phi(t)$$

TABLE I

t	$\Phi(t)$	t	$\Phi(t)$	t	$\Phi(t)$
0·00	0·0000	0·65	0·4843	1·30	0·8064
	399		318		166
0·05	0·0399	0·70	0·5161	1·35	0·8230
	398		306		155
0·10	0·0797	0·75	0·5467	1·40	0·8385
	395		296		144
0·15	0·1192	0·80	0·5763	1·45	0·8529
	393		284		135
0·20	0·1585	0·85	0·6047	1·50	0·8664
	389		272		125
0·25	0·1974	0·90	0·6319	1·55	0·8789
	384		260		115
0·30	0·2358	0·95	0·6579	1·60	0·8904
	379		248		107
0·35	0·2737	1·00	0·6827	1·65	0·9011
	371		236		98
0·40	0·3108	1·05	0·7063	1·70	0·9109
	365		224		90
0·45	0·3473	1·10	0·7287	1·75	0·9199
	356		212		82
0·50	0·3929	1·15	0·7499	1·80	0·9281
	348		200		76
0·55	0·4177	1·20	0·7699	1·85	0·9357
	338		188		69
0·60	0·4515	1·25	0·7887	1·90	0·9426
	328		177		62

TABLE I (contd.)

t	$\Phi(t)$		t	$\Phi(t)$		t	$\Phi(t)$	
1·95	0·9488		2·30	0·9786		2·65	0·9920	
		57			26			11
2·00	0·9545		2·35	0·9812		2·70	0·9931	
		51			24			9
2·05	0·9596		2·40	0·9836		2·75	0·9940	
		47			21			9
2·10	0·9643		2·45	0·9857		2·80	0·9949	
		41			19			7
2·15	0·9684		2·50	0·9876		2·85	0·9956	
		38			16			7
2·20	0·9722		2·55	0·9892		2·90	0·9963	
		34			15			5
2·25	0·9756		2·60	0·9907		2·95	0·9968	
		30			13			5
						3·00	0·9973	

TABLE II

t	$\Phi(t)$	$1 - \Phi(t)$	t	$\Phi(t)$	$1 - \Phi(t)$
1·960	0·95	0·05	2·878	0·996	0·004
2·054	0·96	0·04	2·968	0·997	0·003
2·170	0·97	0·03	3·090	0·998	0·002
2·326	0·98	0·02	3·291	0·999	0·001
2·576	0·99	0·01	3·481	0·9995	0·0005
2·612	0·991	0·009	3·891	0·9999	0·0001
2·652	0·992	0·008	4·417		10^{-5}
2·748	0·994	0·006	4·892		10^{-6}
2·807	0·995	0·005	5·327		10^{-7}

Explanation of Tables I and II, see pp. 36 and 95–96.

VALUES OF THE NORMAL PROBABILITY DENSITY

$$\varphi(z) = \frac{1}{\sqrt{(2\pi)}} e^{-z^2/2}$$

TABLE III

z	Hundredth parts for z									
	0	1	2	3	4	5	6	7	8	9
0·0	$3989 \cdot 10^{-4}$	3989	3989	3988	3986	3984	3982	3980	3977	3973
0·1	3970	3965	3961	3956	3951	3945	3939	3932	3925	3918
0·2	3910	3902	3894	3885	3876	3867	3857	3847	3836	3825
0·3	3814	3802	3790	3778	3765	3752	3739	3726	3712	3697
0·4	3683	3668	3653	3637	3621	3605	3589	3572	3555	3538
0·5	3521	3503	3485	3467	3448	3429	3410	3391	3372	3352
0·6	3332	3312	3292	3271	3251	3230	3209	3187	3166	3144
0·7	3123	3101	3079	3056	3034	3011	2989	2966	2943	2920
0·8	2897	2874	2850	2827	2803	2780	2756	2732	2709	2685
0·9	2661	2637	2613	2589	2565	2541	2516	2492	2468	2444
1·0	2420	2396	2371	2347	2323	2299	2275	2251	2227	2203
1·1	2179	2155	2131	2107	2083	2059	2036	2012	1989	1955
1·2	1942	1919	1895	1872	1849	1826	1804	1781	1758	1736
1·3	1714	1691	1669	1647	1626	1604	1582	1561	1539	1518
1·4	1497	1476	1456	1435	1415	1394	1374	1354	1334	1315
1·5	1295	1276	1257	1238	1219	1200	1182	1163	1145	1127
1·6	1109	1092	1074	1057	1040	1023	1006	9893*	9723*	9566*
1·7	$9405 \cdot 10^{-5}$	9246	9089	8933	8780	8628	8478	8329	8183	8038
1·8	7895	7754	7614	7477	7341	7206	7074	6943	6814	6687
1·9	6562	6438	6316	6195	6077	5959	5844	5730	5618	5508
2·0	5399	5292	5186	5082	4980	4879	4780	4682	4586	4491
2·1	4398	4307	4217	4128	4041	3955	3871	3788	3706	3626
2·2	3547	3470	3394	3319	3246	3174	3103	3034	2965	2898
2·3	2883	2768	2705	2643	2582	2522	2463	2406	2349	2294
2·4	2239	2186	2134	2083	2033	1984	1936	1888	1842	1797
2·5	1753	1709	1667	1625	1585	1545	1506	1468	1431	1394
2·6	1358	1323	1289	1256	1223	1191	1160	1130	1100	1071
2·7	1042	1014	9871*	9606*	9347*	9094*	8846*	8605*	8370*	8140*
2·8	$7915 \cdot 10^{-5}$	7697	7483	7274	7071	6873	6679	6491	6307	6127
2·9	5953	5782	5616	5454	5296	5143	4993	4847	4705	4567
3·0	4432	4301	4173	4049	3928	3810	3695	3584	3475	3370
3·1	3267	3167	3070	2975	2884	2794	2707	2623	2541	2461
3·2	2384	2309	2236	2165	2096	2029	1964	1901	1840	1780
3·3	$1723 \cdot 10^{-6}$	1667	1612	1560	1508	1459	1411	1364	1319	1275

TABLE III (contd.)

z	Hundredth parts for z									
	0	1	2	3	4	5	6	7	8	9
3·4	$1232 \cdot 10^{-6}$	1191	1151	1112	1075	1038	1003	9689*	9358*	9087*
3·5	$8727 \cdot 10^{-7}$	8426	8135	7853	7581	7317	7061	6814	6575	6343
3·6	6119	5902	5693	5490	5294	5105	4921	4744	4573	4408
3·7	4248	4093	3944	3800	3661	3526	3396	3271	3149	3032
3·8	2919	2810	2705	2604	2506	2411	2320	2232	2147	2065
3·9	1937	1910	1837	1766	1698	1633	1569	1508	1449	1393
4·0	1338	1286	1235	1186	1140	1094	1051	1009	9687*	9299*
4·1	$8926 \cdot 10^{-8}$	8567	8222	7890	7570	7263	6967	6683	6410	6147
4·2	5894	5662	5418	5194	4979	4772	4573	4382	4199	4023
4·3	3854	3691	3535	3386	3242	3104	2972	2845	2723	2606
4·4	2494	2387	2284	2185	2090	1999	1912	1829	1749	1672
4·5	1598	1528	1461	1396	1334	1275	1218	1164	1112	1062
4·6	1014	9684*	9248*	8830*	8430*	8047*	7681*	7331*	6996*	6676*
4·7	$6370 \cdot 10^{-9}$	6077	5797	5530	5274	5030	4796	4573	4360	4156
4·8	3961	3775	3598	3428	3267	3112	2965	2824	2690	2561
4·9	$2439 \cdot 10^{-9}$	2322	2211	2105	2003	1907	1814	1727	1643	1563

An asterisk against a number signifies that the number must be multiplied by the power of 10 given in the following row.

THE POISSON DISTRIBUTION

$$P(m, a) = \frac{a^m e^{-a}}{m!}$$

TABLE IV

m \ a	0·1	0·2	0·3	0·4	0·5	0·6
0	0·904837	0·818731	0·740818	0·670320	0·606531	0·548812
1	090484	163746	222245	268128	303265	329287
2	004524	016375	033337	053626	075816	098786
3	000151	001092	003334	007150	012636	019757
4	000004	000055	000250	000715	001580	002964
5		000002	000015	000057	000158	000356
6			000001	000004	000013	000036
7					000001	000003

TABLE IV (contd.)

m \ a	0·7	0·8	0·9	1·0	2	3
0	0·496585	0·449329	0·406570	0·367879	0·135335	0·049787
1	347610	359463	365913	367879	270671	149361
2	121663	143785	164661	183940	270671	224042
3	028388	038343	049398	061313	180447	224042
4	004968	007669	011115	015328	090224	168031
5	000081	001227	002001	003066	036089	100819
6	000008	000164	000300	000511	012030	050409
7	000001	000019	000039	000073	003437	021604
8		000002	000004	000009	000899	008102
9				000001	000191	002701
10					000038	000810
11					000007	000221
12					000001	000055
13						000013
14						000003
15						000001

m \ a	4	5	6	7	8	9
0	0·018316	0·006738	0·002479	0·000912	0·000335	0·000123
1	073263	033690	014873	006383	002684	001111
2	141525	084224	044618	022341	010735	004998
3	195367	140374	089235	052129	028626	014994
4	195367	175467	133853	091226	057252	033737
5	156293	175467	160623	127717	091604	060727
6	104194	146223	160623	149003	122138	091090
7	059540	104445	137677	149003	139587	117116
8	029770	064278	103258	130377	139587	131756
9	013231	036266	068838	101405	124077	131756
10	005292	018133	041303	070983	099262	118580
11	001925	008242	022529	045171	072190	097020
12	000642	003434	011262	026350	048127	072765
13	000197	001321	005199	014188	029616	050376
14	000056	000472	002228	007094	016924	032384
15	000015	000157	000891	003311	009026	019431
16	000004	000049	000334	001448	004513	010930
17	0·000001	000014	000118	000596	002124	005786
18		000004	000039	000232	000944	002893
19		0·000001	000012	000085	000397	001370
20			000004	000030	000159	000617
21			0·000001	000010	000061	000264
22				000003	000022	000108
23				0·000001	000008	000042
24					000003	000016
25					0·000001	000006
26						000002
27						0·000001

INDEX

MADE IN GREAT BRITAIN